T0258124

Encyclopedia of Recrystallization: Minerals and Pharmacology

Volume II

Encyclopedia of Recrystallization: Minerals and Pharmacology
Volume II

Edited by **Sylvia Dickey**

New York

Published by NY Research Press,
23 West, 55th Street, Suite 816,
New York, NY 10019, USA
www.nyresearchpress.com

Encyclopedia of Recrystallization: Minerals and Pharmacology
Volume II
Edited by Sylvia Dickey

International Standard Book Number: 978-1-63238-164-4 (Hardback)

Contents

Preface

This book is a compilation of researches conducted by scientists on issues related to recrystallization. Various disciplines of science, like geology, metallurgy etc., in which recrystallization plays an essential role have been covered. The phenomenon of recrystallization has seen constant growth in the last few decades due to the researches being conducted in various branches of science. This book offers an extensive analysis of recrystallization of minerals and recrystallization in pharmacology, and also discusses proper techniques and methods of its characterization. The book includes the advancements made in diagnostic equipment, and modeling recrystallization.

This book is a comprehensive compilation of works of different researchers from varied parts of the world. It includes valuable experiences of the researchers with the sole objective of providing the readers (learners) with a proper knowledge of the concerned field. This book will be beneficial in evoking inspiration and enhancing the knowledge of the interested readers.

In the end, I would like to extend my heartiest thanks to the authors who worked with great determination on their chapters. I also appreciate the publisher's support in the course of the book. I would also like to deeply acknowledge my family who stood by me as a source of inspiration during the project.

Editor

Part 1

Recrystallization of Minerals

Recrystallization of Fe_2O_3 During the Induration of Iron Ore Oxidation Pellets

Guanghui Li, Tao Jiang, Yuanbo Zhang and Zhaokun Tang
Department of Ferrous Metallurgy,
Central South University Changsha, Hunan,
China

1. Introduction

Magnetite and hematite concentrates are the two main raw materials for oxidized pellet production. Magnetite concentrates have more predominance due to the heat release by oxidation during roasting and may reduce energy consumption (Jiang et al., 2008; Li et al., 2009). However, with the continuous consumption of iron resources driven by the rapid development of iron and steel industry, magnetite resources are becoming scarce and so the development of pellet production is restricted to some extent. Thus, it is imperative to make better use of hematite resources to meet the raw material supply for pellet production (Xu, 2001).

In the pelletizing process, the firing of hematite materials leads to the development of pellet strength by oxide crystal bridging, recrystallization of the Fe_2O_3, as well as the formation of a small quantity of slag (Ball et al., 1973).

As for the firing of hematite materials, more heat should be supplied from external sources due to the absence of the exothermic reaction of oxidation of magnetite. So the energy consumption of hematite pellet production is greater than that of magnetite pellets (Jiang et al., 2008). Moreover, it has been shown that hematite pellet has poor roasting properties and do not achieve adequate physical strength until the roasting temperature is higher than 1300°C. Findings show that the hematite particles and pellet structure keep their original shapes if the temperature is below 1200°C. Thus, the size of hematite particles are not enlarged, nor the Fe_2O_3 crystal lattice defects are eliminated until the temperature is higher than 1300°C. At high temperatures, initial connecting bridges are formed between crystal grains and recrystallization of Fe_2O_3 is observed. However, if the roasting temperature is too high (>1350°C), something detrimental would happen as Fe_2O_3 decomposes to Fe_3O_4 expressed as reaction (1), which adversely results in the loss of pellet quality :

$$6Fe_2O_3 \rightarrow 4Fe_3O_4 + O_2 \quad \begin{aligned} \Delta G^\theta &= 140380 - 81.38T(J) \\ \ln P_{O_2} &= -\frac{70649.22}{T} + 40.96 \end{aligned} \quad (1)$$

From the thermodynamic equation of reaction (1), it can be seen that decomposition temperature of Fe_2O_3 increases with increasing oxygen partial pressure. Therefore, excessively high firing temperature and low oxygen partial pressure should be avoided to restrain the decomposition of Fe_2O_3. Thus, it is necessary to maintain at higher roasting temperature for hematite pellet as well as narrower firing temperature range, which makes the operation of firing equipments difficult.

To enhance the induration of hematite pellets, both magnetite-addition and carbon-burdened methods are found to be the favourable techniques in practice. In this chapter, the induration mechanisms of hematite pellet with addition of magnetite concentrate and anthracite powder are revealed by characterization of recrystallization rules of Fe_2O_3 during the oxidization roasting.

2. Fe_2O_3 recrystallization during the firing of mixed hematite/magnetite concentrates pellet

2.1 Materials and methods

2.1.1 Materials

The chemical compositions of iron ore materials and bentonite are shown in Table 1. The size distribution of iron concentrates is shown in Table 2.

Materials	Fe_{total}	FeO	SiO_2	CaO	MgO	Al_2O_3
Hematite	67.60	0.72	1.55	0.15	0.17	1.16
Magnetite	69.31	27.88	1.32	0.23	0.55	1.02
Bentonite	4.38	/	59.05	0.68	1.73	18.72

Table 1. Chemical compositions of materials / %

Materials	+0.075mm	0.0375-0.075mm	-0.0375mm
Hematite	13.50	16.35	70.15
Magnetite	5.93	15.35	78.72

Table 2. Size distribution of iron ore materials / %

2.1.2 Methods

The experimental procedure includes ball preparation, preheating and roasting tests, strength measurement and mineralogical analysis.

For each trial, 5 kg mixed concentrates at the given hematite/magnetite (H/M) ratio was blended with 8% moisture and 0.5% bentonite was used as binder. The green balls were prepared in a disc pelletizer with a diameter of 1000 mm. The green balls of 10-12 mm in diameter were statically dried at 105°C in an electrical furnace for 4 hrs.

Preheating and roasting tests were carried out in an electrically heated horizontal tube furnace with an internal and external diameter of 50 mm and 70 mm respectively. Firstly, the dry balls were put into a corundum crucible and pushed into the preheating zone of the furnace step by step, preheated at the given temperature for a given period. Then the preheated pellets were taken out of the furnace and cooled in the air, or directly pushed forwards into a higher temperature zone for roasting. Finally, the roasted pellets were taken out and naturally cooled in the air.

The compression strength of cooled pellets was measured with an LJ-1000 material experimental machine. An average value of 20 pellets is expressed as the compression strength for each test.

2.2 Effects of H/M ratio on the compression strength of pellet

According to the orthogonal experimental results, the relationship between H/M ratio in pellet and the compression strength of preheated and roasted pellet was investigated and shown in Fig. 1. Experimental conditions are preheating temperature of 900°C and preheating time of 10 min for Fig. 1a, and 900°C preheating temperature, 10 min preheating time, 1275°C roasting temperature and 15 min roasting time for Fig. 1b.

It can be seen from Fig. 1a, that the preheated pellet strength is continuously improved from 190 N/P to 1132 N/P when magnetite ratio is increased from 0 to 100%. The main reason is that magnetite has fine particle size and great specific surface area, and magnetite particles are rapidly oxidized into Fe$_2$O$_3$ grains during the preheated stage; moreover, the atoms on the newborn Fe$_2$O$_3$ grain surface have greater migrating capability than those on the original hematite grains, and consequently the Fe$_2$O$_3$ crystallites are easily formed between particles. It also can be found that the compression strength of roasted pellet increases with magnetite ratio below 70%; above 70%, the strength will be decreased with the addition of magnetite. This may be due to the fact that some of magnetite particles are residual in the core of pellet during the preheating stage, which cannot be completely oxidized in the high temperature roasting stage; the magnetite particles are recrystallized and bonded with the formation of slag phases.

2.3 Firing properties of mixed H/M pellet

2.3.1 Preheating characteristics

2.3.1.1 Compression strength of preheated pellet

The effects of preheating parameters on the compression strength of preheated pellets with different H/M ratios are shown in Fig. 2. It can be seen from Fig. 2a that the strength of pellet with H/M=70:30 reaches 559 N/P when preheated at 900°C for 10 min, while the strength of pellet with H/M=50:50 exceeds 560 N/P when the time is more than 5 min. Fig. 2b shows that the preheated pellet strength keeps increasing with the preheating temperature varying from 800 to 1000°C. The strength of the pellets with H/M=50:50 reaches 470 N/P when the temperature is 800°C. However, the compression strength of the preheated pellets with H/M=70:30 is less than 400 N/P until the temperature is over 850°C. Thus, the results show that the preheating time can be shortened and preheating temperature can be decreased with the increase of magnetite ratio in the pellet.

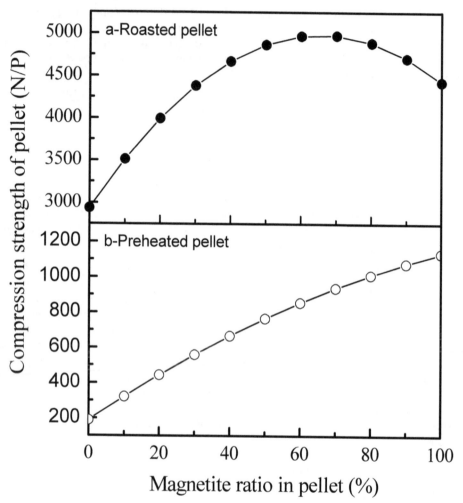

Fig. 1. Effects of magnetite ratio on the compression strength of pellet

2.3.1.2 FeO content of preheated pellet

Whether the magnetite in pellet is oxidized completely or not has an important effect on the roasted pellet performance (compression strength, microstructure, mineral composition, etc.). Generally speaking, residual FeO content in the preheated pellet is required to be less than 3% in operation. The FeO content in pellets with H/M=70:30 under different preheating conditions were analyzed and are shown in Fig. 3. FeO content falls gradually with the increase in preheating temperature and time. FeO content is less than 3% only if the preheating time is more than 11 min at 850°C; however, at 900°C, FeO content goes down to 2.97% when the preheating time is 8 min. In consideration of the compression strength and FeO content together, the suitable preheating parameters for the pellet with H/M=70:30 should be 8~11 min at 850°C~900°C.

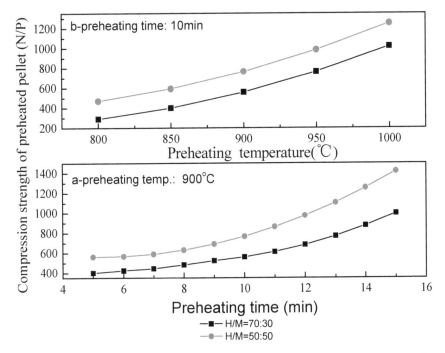

Fig. 2. Effects of preheating on the compression strength of the preheated pellets with different H/M ratios

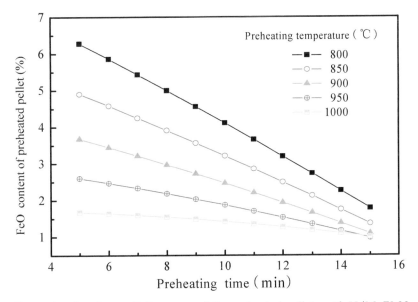

Fig. 3. Effects of preheating on FeO content of the preheated pellets with H/M=70:30

2.3.2 Roasting characteristics

The effects of preheating and roasting parameters on the compression strength of the finished pellets with different H/M ratios are presented in Fig. 4.

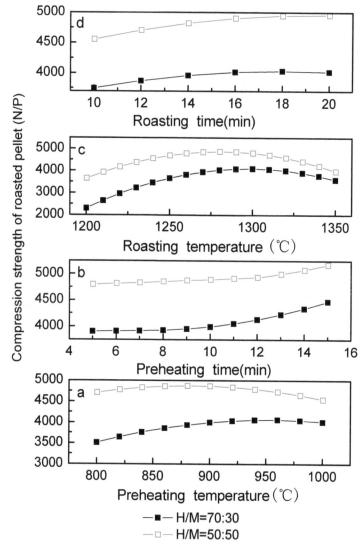

a-preheating time 10 min, roasting temp. 1275°C, roasting time 15 min;
b-preheating temp. 900°C, roasting temp. 1275°C, roasting time 15 min;
c-preheating temp. 900°C, preheating time 10 min, roasting time 15 min;
d-preheating temp. 900°C, preheating time 10 min, roasting temp. 1275°C

Fig. 4. Effects of preheating and roasting parameters on the compression strength of the finished pellet with different H/M ratios

As shown in Fig. 4a, the compression strength of roasted pellets increases quickly with the preheating temperatures varying from 800 to 900°C and remains almost unchanged when temperature is between 900 and 950°C. While the temperature is further increased, the pellet strength decreases slightly. The main reason is that a compact layer is quickly formed and the residual magnetite in the core of pellet is difficult to be oxidized completely. If pellet is roasted in poor oxidative atmosphere at relatively high temperature, low melting point slag phases (such as 2FeO · SiO$_2$) are formed, and result in deteriorating the compression strength of roasted pellet. From Fig. 4b, the strength of the roasted pellets increased gradually with increasing preheating time, the reason for which is that magnetite particles are gradually oxidized into Fe$_2$O$_3$ crystallites and they are recrystallized more completely with the prolonged preheating time.

As shown in Fig. 4c, the pellet strength increases steadily with the increase of roasting temperature below 1300°C. The possible reason is that the Fe$_3$O$_4$ is gradually oxidized into Fe$_2$O$_3$, which leads to Fe^{3+} diffusion, rearrangement of Fe$_2$O$_3$ crystal lattices and a compact microstructure formed. When the roasting temperature is over 1300°C, however, the pellet strength is reduced with rising temperature, the main reason for which is that: on one hand, it is too late for the residual magnetite in the core of preheated pellet to be oxidized adequately at high temperature, so that low melting point slag phases are formed priorly; and on the other hand, part of Fe$_2$O$_3$ crystal grains decomposes at high temperature reversely, so that the structure of the pellet is destroyed to some degree. As far as the roasting time is concerned, as given in Fig. 4d, the pellet strength is significantly improved with the roasting time in the range of 10–15 min and remains constant after 15 min.

The results mentioned above show that adding a proper proportion of magnetite concentrate into the hematite pellet is able to improve the compression strength of both preheated and roasted pellets.

2.4 Induration mechanisms of mixed H/M pellet

Mineral composition, microstructure and Fe$_2$O$_3$ crystallization of fired pellets at various preheating and roasting temperatures were studied by using Leica DMRXE microscope with an automatic image analyzer.

2.4.1 Crystallization behavior of Fe$_2$O$_3$ during preheating

Oxidation of magnetite into hematite is the main reaction during preheating. Oxygen readily diffuses into the interior of the porous pellet and reacts with magnetite particles during preheating. Oxidation always occurs firstly on the surface of particles and cracks, and usually a few small spotty or lamellae hematite are formed (Figs. 5, 6). The oxidation process will advance towards the core with increasing temperature.

It is shown that the hematite in the pellets involves two types: the one is original hematite (OH), which is the unreacted hematite from the raw hematite concentrate (particle A), and the other, namely secondary hematite (SH), is formed from the oxidation of magnetite concentrate (particle B).

Obvious differences in colour and shape can be observed between the OH and SH grains. For the OH grains, their colours are bright and white, and their shapes are original and with discernible angularity. The strip, triangular and rectangular grains can be distinctly observed, and the surface of crystal grain is smooth and the compact inner structure is unchanged. Moreover, the distance between the two close OH particles is large; there is no crystallitic bond formation between them.

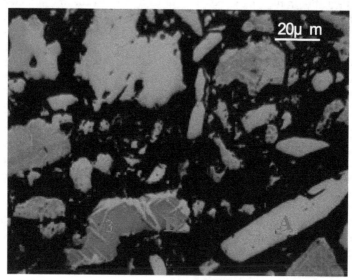

1-SH (columnar); 2-OH; 3-residual magnetite (irregular, tabular)

Fig. 5. Shapes of SH and OH grains in the pellet preheated at 950°C

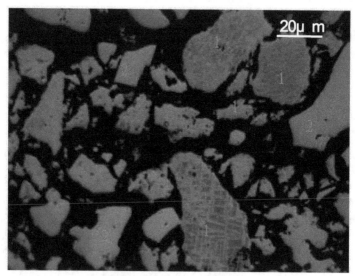

1-SH (reticular); 2-OH; 3-residual magnetite (vein)

Fig. 6. Shapes of SH and OH grains in the pellet preheated at 1000°C

By contrast, for the SH grains, they are relatively heavy-coloured, and some residual magnetite domains can be observed. Tabular or massive grains are the main morphology of residual Fe$_3$O$_4$ preheated at 950°C (Fig.5), but the grains change into reticular and vein shape at 1000°C (Fig. 6).

In comparison with OH grains, the angularities of the SH particles disappear or become unclear during oxidation, and they transform into massive, zonal or columnar-shaped particles. The formation of Fe$_2$O$_3$ microcrystalline junctions between the close SH particles is significantly different from the OH particles at this stage.

The above results suggest that Fe$_2$O$_3$ in SH particles is more active than that in the OH particles. Fe$_2$O$_3$ microcrystalline junctions between SH particles are formed when the pellet is preheated, which is able to improve the strength of the preheated pellet. The OH particles keep their original shapes, and no Fe$_2$O$_3$ microcrystalline junction can be observed at the preheating stage, thus OH has little contribution to the strength of preheated pellets. Therefore, the improvement of oxidative atmosphere during preheating is able to enhance Fe$_2$O$_3$ microcrystalline junctions between the SH particles, the strength of preheated pellet will be improved accordingly.

2.4.2 Crystallization behaviour of Fe$_2$O$_3$ during roasting

The crystal morphology of Fe$_2$O$_3$ and microstructure of pellets roasted at different temperatures are shown in Figs. 7–9.

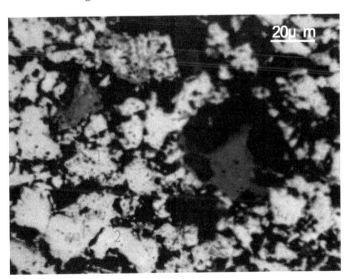

1-SH particle; 2-OH particle

Fig. 7. OH grains enclosed by or connected with SH grains in the pellet roasted at 1150°C

As shown in Fig. 7, a large number of junctions between particles are formed by recrystallization of Fe$_2$O$_3$ at 1150°C and the strength of the roasted pellet highly increases. However, the inner structure and shape of the OH particles remain visible, which indicates that none of Fe_2O_3 recrystallization takes place in the OH particles at 1150°C. The junctions between particles come from Fe_2O_3 crystallization of the SH particles.

1-SH particle; 2-OH particle

Fig. 8. Recrystallization of Fe_2O_3 in the pellet roasted at 1230°C

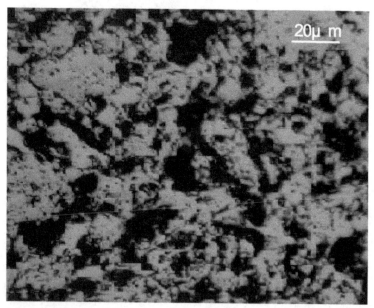

Fig. 9. Developed Fe_2O_3 recrystallization in the pellet roasted at 1280°C

At 1230°C (Fig. 8), Fe$_2$O$_3$ recrystallization is further developed and the interconnection between particles has been enhanced, and individual particles are scarce in pellet. The OH particles have been connected with or even enclosed by SH particles, and the strength of pellet is further increased. However, the inner structure of the OH particles are still compact, with marked difference from the SH particles, which suggests that recrystallization of Fe$_2$O$_3$ in the OH particles is still undeveloped at 1230°C, and the junctions between particles are from the Fe$_2$O$_3$ recrystallization of SH particles mainly.

At 1280°C, it can be seen from Fig. 9 that Fe$_2$O$_3$ recrystallization in the OH particles develops satisfactorily; all of hematite particles become porous, the profile of the OH particles almost disappears, and all the particles (both SH and OH) are connected with each other through Fe$_2$O$_3$ recrystallization and form a whole crystal structure; therefore, the strength of pellet is further improved.

In summary, the results indicate that Fe$_2$O$_3$ from SH and OH particles have different activities during roasting, which results in the difference of strength formation mechanisms of the mixed H/M concentrates pellets at various roasting temperatures. At lower temperature (1150°C), the pellet strength is mainly provided by Fe$_2$O$_3$ recrystallization of the SH particles. However, OH particles can be connected with the SH particles through crystal bridges formed by high activity Fe$_2$O$_3$ of SH particles, or even enclosed by the SH particles. So, OH particles also contribute to the pellet strength to a certain degree.

When the temperature goes up to 1250°C, Fe$_2$O$_3$ recrystallization in the SH particles is further developed, and the crystal junctions between the particles become stronger. However, Fe$_2$O$_3$ recrystallization within OH particles does not obviously take place, and few joint is formed between the close OH particles. Pellet strength is mainly provided by Fe$_2$O$_3$ recrystallization junctions of SH particles.

At 1280°C, Fe$_2$O$_3$ recrystallization in the SH and OH particles simultaneously occurs, and the joints between the close particles, including OH and SH, are well developed. Particles are connected with each other and the roasted pellet forms a whole crystal structure. The Fe$_2$O$_3$ recrystallization of SH and OH particles plays a crucial role in improving the roasted pellet strength.

The results indicate that, because the activity of Fe$_2$O$_3$ in the SH grains is higher than that in the OH grains, Fe$_2$O$_3$ recrystallization bonds between particles can be enhanced by the SH grains during preheating and roasting, thus the newborn SH is able to improve the pellet strength and decrease the roasting temperature of mixed H/M concentrates pellet. It is the reason that adding a certain proportion of magnetite concentrate is for the enhanced roasting performance of hematite pellet.

3. Fe$_2$O$_3$ recrystallization during the firing of carbon-burdened hematite pellet

3.1 Materials and methods

3.1.1 Materials

The hematite concentrate is characterized by high total iron grade (67.2% TFe) and less impurities (Table 3). The particle size is 92% undersize 0.074mm and the specific surface area is 1629.5cm^2/g (Table 4).

Anthracite powder was used as the material of burdened carbon, of which the specific surface area reaches 6599 cm²/g after grinding.

Materials	Fe_{total}	FeO	SiO_2	CaO	MgO	Al_2O_3	LOI*
Hematite	67.22	0.55	2.17	0.01	0.05	0.55	0.59
Anthracite	1.48	/	6.86	1.51	0.24	5.88	84.08
Bentonite	7.07	/	60.61	0.94	2.2	17.98	10.41

*LOI-Loss on ignition

Table 3. Chemical compositions of the materials /%

+0.074mm/%	0.074-0.045mm/%	-0.045mm/%	specific surface area/cm².g⁻¹
7.79	28.99	63.22	1629.5

Table 4. Size distribution and specific surface area of hematite concentrate

3.1.2 Methods

For each test, 5 kg of hematite concentrate was blended with the given proportion anthracite powder, using 1.25% bentonite as binder. The green balls were prepared in a disc pelletizer with a diameter of 1000 mm, and the green balls with 9~15 mm in diameter were statically dried at 105°C in a drying oven for 4 hours.

Firing teats were carried out in an electrically heated shaft furnace. To simulate firing atmosphere, mixed gas of N_2/O_2 at the given oxygen partial pressure (volume fraction) was pumped into the shaft furnace at a certain flow-rate. The dry balls were charged into heat-resistant pot, and then the pot was pushed downwards into the furnace step by step. The pellets were fired at the given temperature for a given period. Afterwards, the roasted pellets were taken out and naturally cooled to ambient temperature. The compression strength of cooled pellets was measured with an LJ-1000 material experimental machine. An average value of 10 pellets is expressed as the compression strength for each test.

3.2 Firing characteristics of carbon-burdened hematite pellet

3.2.1 Effects of anthracite dosage on compression strength

The effects of anthracite dosage on the compression strength of the roasted pellets are shown in Fig. 10.

As shown in Fig. 10, the pellet strength with 0.5% anthracite is a little lower than that with no anthracite. When anthracite dosage reaches 1.0%~1.25%, the compression strength goes up to the maximum, however, the strength decreases greatly if the anthracite further increases from 1.5% to 4%. The results indicate appropriate anthracite amount of 1.0%~1.25% may improve the strength of hematite pellet.

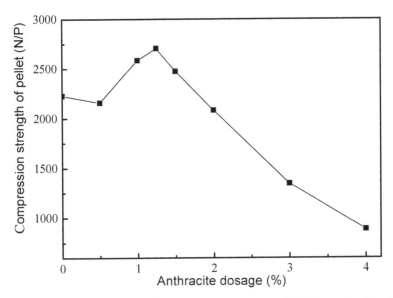

(Oxygen partial pressure: 20%, airflow: 6L/min, roasting temperature: 1280°C, roasting time: 20min)

Fig. 10. Effects of anthracite dosage on the compression strength of pellet

3.2.2 Effects of roasting temperature on compression strength

The effects of roasting temperature on compression strength of pellet are shown in Fig. 11.

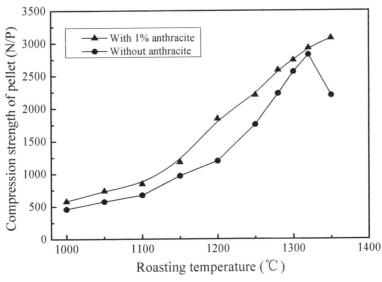

(Oxygen partial pressure 20%, airflow: 6L/min, roasting time: 20min)

Fig. 11. Effects of roasting temperature on the compression strength of pellet

As shown in Fig.11, the strength of carbon-burdened pellet is always higher than that of pellet without carbon when roasted at the same temperature; moreover, for carbon-burdened hematite pellet, the compression strength increases constantly with the temperature. However, the strength of pellet in the absence of carbon not only doesn't increase markedly until 1250°C, but also decreases over 1320°C due to the decomposition of Fe_2O_3. The results imply that the roasting temperature can be decreased and the firing temperature range is enlarged by adding an appropriate amount of anthracite into hematite pellet.

3.2.3 Effects of roasting time on compression strength

The compression strength of roasted pellet increases gradually with the roasting time prolonging and a maximum strength is obtained at 25 min (in Fig.12), hereafter, the strength almost remains constant.

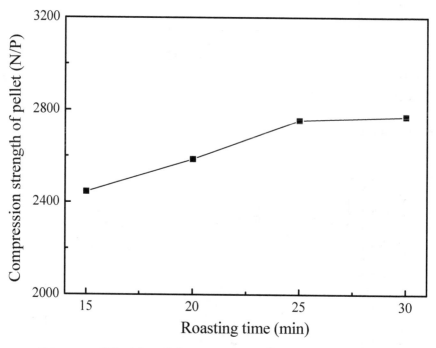

(Oxygen partial pressure: 20%, airflow: 6L/min, roasting temperature: 1280°C, anthracite: 1.0%)

Fig. 12. Effects of roasting time on the compression strength of carbon-burdened pellet

3.2.4 Effects of oxygen partial pressure on compression strength

The compression strength of pellet is susceptible to the change of oxygen partial pressure as shown in Fig.13. The strength of the pellet roasted in N_2, that is, in the inert atmosphere, is slightly higher than that in 10% O_2. The strength nearly reaches the maximum at 20% O_2, and then decreases gradually with increasing the oxygen partial pressure.

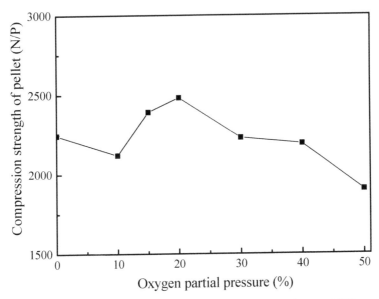

(Airflow: 6L/min, roasting temperature: 1280°C, roasting time: 20min, anthracite: 1.0%)

Fig. 13. Effects of oxygen partial pressure on the compression strength of carbon-burdened pellet

3.3 Roles of burdened carbon during the roasting

To shed light on the effects of the burdened carbon on the induration of hematite pellet, a test, as shown in Fig. 14, was designed to identify the reduction and decomposition of hematite in carbon-burdened pellet during roasting.

As shown in Fig. 14, a cylinder was made by briquetting hematite concentrate firstly, the cylinder bottom is closed, and its inner diameter is 20 mm, the outer diameter is 30 mm. To allow the gas upward injecting into the inner cylinder and penetrating through the anthracite powder layer, many ventages with 0.1 mm diameter were drilled through the cylinder bottom.

Dry pellets with 2~3 mm in diameter were prepared from hematite concentrate in advance, and then were charged into the surface layer of inner cylinder. The cylinder bottom, anthracite powder layer and pellet layer were separated by inert material of Al_2O_3 powder to avoid their contact with each other.

At the beginning of trail, the sample prepared according to Fig. 14 was placed in an electrically heated shaft furnace, and 6L/min N_2 with 99.99% purity was pumped into the shaft furnace from the bottom. The sample was taken out and immersed into water immediately after roasted at 1280°C for 20 minutes. Subsequently, the FeO content of the cylinder bottom and pellet was measured respectively.

It is shown that FeO content of the cylinder bottom and pellet is 4.35% and 28.69% respectively. FeO content of the pellet is obviously higher than that of the cylinder bottom.

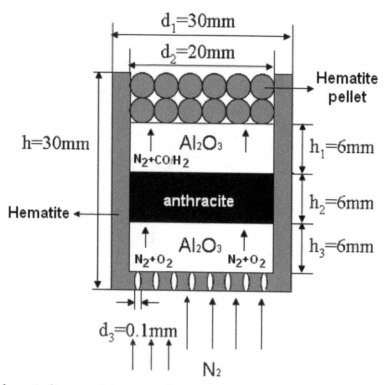

Fig. 14. Schematic diagram of the test on the role of burdened carbon during roasting

In N_2 atmosphere, hematite may decompose into magnetite and release O_2 according to formula (1), and FeO content increases accordingly. It is the reaction that the hematite within the cylinder bottom occurs. However, because of being separated by Al_2O_3 powder, the hematite within the cylinder bottom can't be reduced by anthracite or upwards flowing reductive gases CO/H_2, which are produced by gasification of anthracite. Therefore, the increase of FeO content in the cylinder bottom is only caused by the decomposition of hematite.

However, as regards as the hematite within the pellet, on one hand, it may be decomposed into magnetite in N_2 gas as same as the hematite within the cylinder bottom; on the other hand, it can be also reduced into magnetite by upwards flowing CO/H_2 produced by gasification of anthracite. Therefore, the increase of FeO content of the pellet is caused by both the decomposition and the reduction of hematite, and the latter is more crucial.

The above results identify that the burdened carbon plays the role of reductant during the roasting, and a large number of newborn magnetite are created due to the reductive reaction of hematite by CO/H_2, the products of gasification of anthracite.

Of cause, the burdened carbon can release heat via combustion and heat up the inner pellet, which is advantageous for the induration of hematite pellet. The heating function of the burdened carbon resembles the heat release by oxidation of the magnetite concentrate added into hematite pellet.

3.4 Induration mechanisms of carbon-burdened hematite pellet

3.4.1 Changes of FeO content during the roasting of carbon-burdened pellet

3.4.1.1 Effects of anthracite dosage

FeO content of the roasted pellets with different dosage of anthracite is measured by chemical analysis and the results are shown in Fig. 15.

As shown in Fig.15, the FeO content of the roasted pellets increases gradually with the increasing the dosage of anthracite. This suggests that the anthracite in pellet benefits the reduction of hematite during roasting. Because of the relatively strong reductive atmosphere under the condition of high dosage of anthracite, a large amount of Fe$_2$O$_3$ is reduced into Fe$_3$O$_4$.

The above results that FeO content of the pellet varies with the anthracite dosage are in accord with the conclusions obtained from the tests as shown in Fig. 14

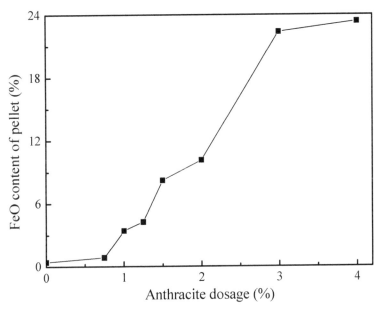

(Oxygen partial pressure: 20%, airflow: 6L/min, roasting temperature: 1280°C, roasting time: 6min)

Fig. 15. Effects of anthracite dosage on FeO content of the carbon-burdened pellet

3.4.1.2 Effects of oxygen partial pressure

FeO content of the pellets roasted at different oxygen contents is measured and the results are shown in Fig. 16.

As shown in Fig.16, the reduction of hematite mainly occurs at the initial roasting stage. The time, when the maximum FeO content attains to, is shortened with increasing the oxygen partial pressure, which indicates that the increase of oxygen partial pressure has favourable effect on oxidization rate of newborn magnetite.

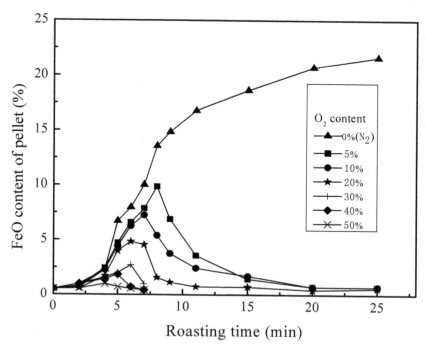

(Anthracite: 1.0%, airflow: 6L/min, roasting temperature: 1280°C)

Fig. 16. Change of FeO content of the pellets roasted at different oxygen partial pressures for different period

The maximum FeO content decreases when the carbon-burdened hematite pellet is roasted at high oxygen partial pressure. The result shows that the lower the oxygen partial pressure during roasting, the stronger the reductive atmosphere within the pellet; and the reduction of Fe_2O_3 can be enhanced. At high oxygen partial pressure, the burdened carbon in pellet combusts completely, and more carbon comes into being not CO but CO_2, thus reductive atmosphere is weakened, which prevents the hematite from reducing into magnetite.

In oxidative atmosphere, FeO content of the pellet increases firstly and then decreases during roasting, the reason for which is that, at the initial roasting stage, the reduction rate of hematite into magnetite is higher than the oxidation rate of newborn magnetite into hematite, however, accompanied with the consumption of burdened carbon, the reduction rate decreases while the oxidation rate increases, the maximum FeO content attains when the reduction rate equals to the oxidation rate. Subsequently, FeO content decreases alone with the oxidation of magnetite untill the oxidation is complete.

It indicates that some original hematite (OH) can be reduced to magnetite by the burdened carbon firstly, and then the newborn magnetite is oxidized into secondary hematite (SH) again during the roasting. The transformation of OH to magnetite and newborn magnetite to SH changes the route of Fe_2O_3 recrystallization during the induration of carbon-burdened hematite pellet.

3.4.2 Fe$_2$O$_3$ recrystallization behaviours of carbon-burdened hematite pellet

To shed light on the Fe$_2$O$_3$ recrystallization behaviour, the microstructure of carbon-burdened hematite pellets at different roasting stage is investigated and the results are presented in Fig. 17.

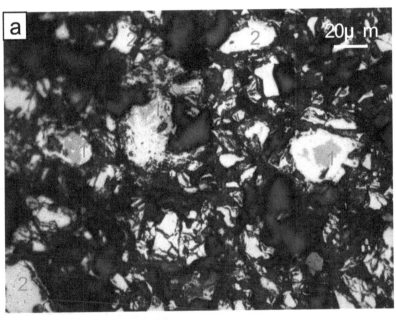

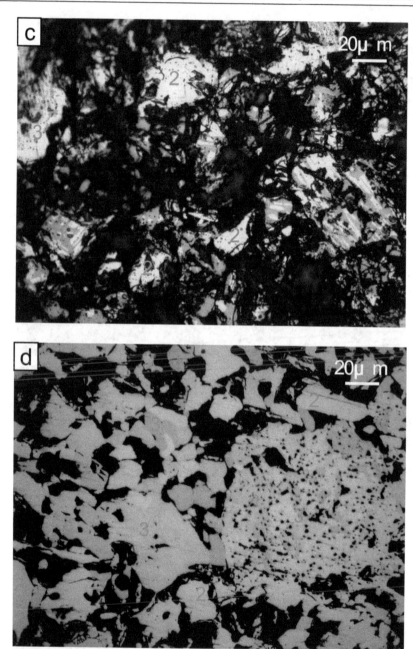

Roasting time: a–4min; b-6min; c-8min; d-20min
1-Magnetite; 2-OH ; 3- SH
(Anthracite: 1.0%, Oxygen partial pressure: 20%, airflow: 6L/min)

Fig. 17. Microstructure of the carbon-burdened pellets roasted at 1280°C

When the volatile matter in anthracite is pyrolyzed and gasified during initial roasting stage, the reduction atmosphere is gradually enhanced within the pellet, and a few magnetite grains, shown in Fig. 17a, are formed by the reduction of OH. An apparent colour difference can be observed between the hematite grains and magnetite grains. The hematite presents bright and white colour, while the magnetite is grey. However, the OH particles keep their original shape and discernible angularity. The inner crystal structure of OH is compact and there is no crystal bond formed between OH particles.

It can be seen from Fig. 17b, a large amount of newly created magnetite is observed as the roasting process progresses; however, the OH particles keep their original shapes, and there is no crystal junction observed still.

As shown in Fig. 17c, the content of newborn magnetite decreases, and a few SH, which comes from the oxidation of the newborn magnetite, are formed. In comparison with OH grains, the angularities of the SH particles become unclear, and a few crystal junctions can also be observed between them. In this stage, with the consumption of anthracite, reduction atmosphere is weakened and oxidation rate of SH is accelerated.

At the end of roasting, the newborn magnetite has been completely oxidized into SH, and a large number of crystal junctions between particles are formed (as shown in Fig. 17d). Pellet strength is mainly provided by Fe$_2$O$_3$ recrystallization junctions between SH particles and OH particles.

It can be concluded from the results mentioned above that partial OH particles can be reduced firstly and turn into magnetite particles by the anthracite powder dispersed in the carbon-burdened hematite pellet, however, the newborn magnetite can be subsequently oxidized into SH particles with higher activity, and the route of Fe$_2$O$_3$ recrystallization is changed from the recrystallization of OH particles to the recrystallization bonds among SH particles and OH particles. Therefore, the formation of SH during the roasting of carbon-burdened hematite pellet is able to improve the pellet strength and decrease the roasting temperature. It is the reason why adding a certain proportion anthracite is also an effective way to enhance the roasting performance of hematite pellet.

4. Conclusions

Both magnetite and burdened carbon are found to be the favourable techniques for enhancing the induration of hematite pellet. The induration mechanisms of hematite pellet with addition of magnetite concentrate and anthracite powder are revealed by characterization of Fe$_2$O$_3$ recrystallization rules during the oxidization roasting.

The crystallization behaviours of Fe$_2$O$_3$ during preheating and roasting of pellets made from mixed hematite/magnetite (H/M) concentrates have been revealed. The results indicate that the strength of pellet is mainly provided by the crystalline connections between the particles during preheating. This occurs because the activity of Fe$_2$O$_3$ from secondary hematite (oxidized from magnetite concentrate) is higher than that from original hematite (from the raw hematite concentrate). In the roasting process, when temperature is lower than 1250°C, the strength is mainly provided by the development, connection and growth of Fe$_2$O$_3$ crystalline grains from secondary hematite. Only if the temperature exceeds 1280°C does Fe$_2$O$_3$ recrystallization in original hematite grains develop very well.

Because the activity of Fe_2O_3 in the secondary hematite grains is higher than that in the original hematite grains, Fe_2O_3 recrystallization bonds between particles can be enhanced by the secondary hematite grains during preheating and roasting, and the secondary hematite in H/M concentrate pellet is able to improve the strength and decrease the roasting temperature of hematite pellet. Thus, adding a certain proportion of magnetite concentrate is an effective way to improve the roasting performance of hematite pellet.

The effects of anthracite on oxidation roasting behaviour for hematite pellet have been elucidated. Anthracite in pellet has two functions: the one lies on that a part of heat needed in roasting process can be supplied by the carbon combustion, and the other is that the reduction of partial hematite by the carbon dispersed in pellet, as well as the partial decomposition of hematite at relatively low oxygen partial pressure, leads to the transformation of hematite into magnetite during the roasting.

Based on microstructure analysis, it can be founded that the new-born magnetite, produced from the reduction and the decomposition of original hematite (OH), is oxidized into the secondary hematite (SH) during roasting. Thus, Fe_2O_3 recrystallization bonds between particles can be consolidated by the secondary hematite grains at lower roasting temperature. The strength of carbon-burdened hematite pellet is enhanced and the roasting temperature is decreased due to the formation of secondary hematite. Therefore, adding a certain proportion of anthracite is also an effective way to enhance the roasting performance of hematite pellet.

5. Acknowledgments

The authors want to express their thanks to National Science Fund for Distinguished Young Scholars (50725416), National Natural Science Foundation of China (50604015 and 50804059) and Fundamental Research Funds for the Central Universities for financial supports of this research. Dr. Mingjun Rao is appreciated for his helpful remarks on spelling and expression.

6. References

Jiang, T.; Zhang, Y. & Huang, Z. (2008). Preheating and Roasting Characteristics of Hematite–Magnetite (H–M) Concentrate Pellets. *Ironmaking Steelmaking*, Vol. 35, No. 1, pp. (21-26), ISSN 0301-9233

Li, G.; Li, X. & Zhang, Y. (2009). Induration Mechanisms of Oxidised Pellets Prepared from Mixed Magnetite–Haematite Concentrates. *Ironmaking Steelmaking*, Vol. 36, No. 5, pp. (393-396), ISSN 0301-9233

Xu, M. (2001). Development of BF Burden Structure in China. *Ironmaking*, Vol. 20, No. 2, pp. (24-27), ISSN 1001-1471

Ball, D.; Dartnell, J. & Davison, J. (1973). *Agglomeration of Iron Ores*, Heinemann Educational, ISBN 0435720104, London

APbill, J. (1985). Carbonaceous Additives for Pelletizing Production, 4TH *International Symposium on Agglomeration*, ISBN 0-932897-00-2, Toronto, Canada, June, 1985

Clout, J.; Manuel J. (2003). Fundamental Investigations of Differences in Bonding Mechanisms in Iron Ore Sinter Formed from Magnetite Concentrates and Hematite Ores. *Powder Technology*, Vol. 130, pp. (393-399), ISSN 0921-8831

Yang, X.; Guo, Z. & Wang, D. (1995). Research on the Reduction Mechanism of Iron Ore Pellets Containing Graphite. *Engineering Chemistry and Metallurgy*, Vol. 16, No. 2, pp. (118-126), ISSN 1001-2052

2

Steady-State Grain Size in Dynamic Recrystallization of Minerals

Ichiko Shimizu
Department of Earth and Planetary Science, University of Tokyo, Tokyo
Japan

1. Introduction

Dynamic recrystallization (DRX) is a strain restoration and grain refinement mechanism that occurs in high-temperature dislocation creep of metals and minerals (Humphreys & Hatherly, 2004). Microstructures indicative of DRX are commonly observed in rock-forming minerals that have been subjected to natural deformation in the Earth's crust and mantle (Fig. 1).

Laboratory studies have revealed that the average size d of recrystallized grains approaches a steady-state value, which is determined by the applied stress and is independent of the initial grain size. Twiss (1977) proposed a stress–grain size relation of the following form:

$$\frac{d}{b} = K \left(\frac{\sigma}{\mu}\right)^{-p} \tag{1}$$

where σ is the flow stress, μ is the shear modulus, b is the length of the Burgers vector, and K is a non-dimensional constant. The grain size exponent p ranges between 1 and 1.5 for most materials. Empirically determined σ–d relations of minerals have been used to estimate the stress states in the Earth's interior. However, detailed studies of a Mg alloy (De Bresser et al., 1998) and NaCl (Ter Heege et al., 2005) revealed that K has a weak dependence on temperature. Derby & Ashby (1987) modeled the DRX processes of metals and predicted the temperature dependence of the recrystallized grain size, but they failed to account for the observed range of exponent p (Derby, 1992; Shimizu, 2011).

In this chapter, we focus on deformation and recrystallization processes in minerals and examine the effects of stress and temperature on the steady-state grain size.

2. Recrystallization mechanisms in minerals

DRX was first observed in hot deformation of cubic metals such as Cu, Ni, and austenitic iron. A simplified description of DRX in these metals is as follows. Strain-free new grains are usually formed by bulging of pre-existing grain boundaries and they grow at the expense of old grains to reduce the dislocation energy of the material (Sakai, 1989; Sakai & Jones, 1984). As the dislocation density of the new grains increases, they cease to grow and new nucleation events occur at their margins. These processes repeat cyclically during dislocation creep.

Fig. 1. Optical micrograph of a thin section of a quartz schist (Sanbagawa metamorphic belt, Japan) under polarized transmitted light with a sensitive color plate. Blue and red represent the orientation of the crystallographic c-axis of quartz grains. New small grains form at the margins and interiors of larger grains.

In contrast to the classical view of DRX described above, syndeformational recrystallization of minerals such as quartz, calcite, and olivine proceeds with progressive misorientation of subgrain boundaries (Poirier, 1985). Subgrain rotation (SGR) recrystallization also occurs in some metals such as Mg and Al alloys and is termed continuous DRX, whereas DRX in the original sense is currently referred to as discontinuous DRX (Humphreys & Hatherly, 2004). At low temperatures (T) and high strain rates ($\dot{\varepsilon}$), SGR is localized at grain margins (Hirth & Tullis, 1992; Schmid et al., 1980); however, intracrystalline SGR becomes more important and grain boundary migration (GBM) occurs at high T and low $\dot{\varepsilon}$ (Hirth & Tullis, 1992; Rutter, 1995) (Fig. 2). Consequently, the recrystallized grain size is much larger than the subgrain size (Guillopé & Poirier, 1979; Karato et al., 1980).

For both discontinuous and continuous DRX, grain size reduction occurs at nucleation events, whereas strain-induced GBM leads to overall coarsening. The steady-state grain size is determined by the dynamic balance between nucleation and grain growth (Derby & Ashby, 1987).

3. Grain size distribution

In the σ–d relation (Eq. 1), the steady-state microstructure is represented by a single value of the 'average' grain size d, but dynamically recrystallized materials generally have wide grain size distributions. As a simplified model of DRX, Shimizu (1998a; 1999; 2003) considered following nucleation and growth processes and analyzed the evolution of the grain size distribution:

1. Nucleation occurs at a constant rate I per unit volume.

2. Nucleation sites are randomly distributed.

3. Each grain grows with a radial growth rate $\dot{R}$.

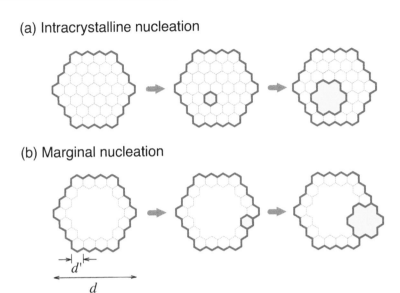

(a) Intracrystalline nucleation

(b) Marginal nucleation

Fig. 2. Nucleation and growth in continuous DRX. Solid lines represent grain boundaries and thin dotted lines represent subgrain boundaries. Nucleated grains (yellow) are formed by SGR and grow in the deformed matrix.

4. Newly crystallized grains replace older grains.

In the steady state, the grain size has a nearly a log-normal distribution and many newly crystallized grains coexist with a few old grains in a certain population balance. The average grain size satisfies

$$d = a \left(\frac{\dot{R}}{I} \right)^{\frac{1}{4}} \tag{2}$$

where a is a scaling factor; $a = 1.14$ for a 3D distribution and $a = 1.12$ for a distribution measured in a 2D section. Shimizu (1998b; 2008; 2011) considered strain-induced grain growth for $\dot{R}$ (Sec. 4) and SGR nucleation for I (Sec. 5) and derived the σ–d relation for continuous DRX (Sec. 6). In Sec. 7, we revise the theoretical model to incorporate the influence of the surface-energy drag.

4. Strain-induced grain growth

4.1 Dislocation energy

During high-T dislocation creep of minerals, dynamic recovery cooperates with continuous DRX and assists subgrain formation. Unrecovered microstructures such as tangled dislocations are rarely observed in recrystallized grains (Hirth & Tullis, 1992). Hence, the strain energy (E_{strain}) is given by a sum of the energies of isolated dislocations and sub-boundaries (E_{disl} and E_{sub}, respectively):

$$E_{strain} = E_{disl} + E_{sub} \tag{3}$$

The free dislocation energy per unit volume is

$$E_{disl} = \rho \zeta \tag{4}$$

where ρ is the dislocation density and ζ is the dislocation line tension. When the internal stress around dislocations is equilibrated with the applied stress σ, the following equation holds (Nabarro, 1987):

$$\sigma = \alpha \mu b \rho^{\frac{1}{2}} \tag{5}$$

where α is a constant that depends on the configuration of the dislocation arrays. Hence,

$$\rho = \left(\frac{\sigma}{\alpha \mu b} \right)^2 \tag{6}$$

The dislocation line tension is given by (Hirth & Lothe, 1982)

$$\zeta = \frac{\mu b^2 \chi}{4\pi} \ln \left(\frac{\beta r}{b} \right) \tag{7}$$

where r is the characteristic radius of the elastic field around the dislocation core and the constant β is typically in the range 3–4. The parameter χ depends on the dislocation configuration:

$$\begin{cases} \chi = 1; & \text{for a screw dislocation} \\ \chi = \dfrac{1}{1-\nu}; & \text{for an edge dislocation} \end{cases} \tag{8}$$

where ν is Poisson's ratio. For a first-order approximation, we assume that all dislocations are edge dislocations. Considering that the elastic field around a dislocation is canceled by other dislocations at half the distance between them, r is scaled as

$$r = \frac{1}{2} \rho^{-\frac{1}{2}} \tag{9}$$

Substituting Eqs. (6) and (9) into Eq. (7) yields

$$\zeta = \frac{\mu b^2}{4\pi(1-\nu)} \ln \left(\frac{\beta \alpha \mu}{2\sigma} \right) \tag{10}$$

Substituting Eq. (10) into Eq. (4) and using Eq. (6) again, we have

$$E_{disl} = \frac{\sigma^2}{4\pi \alpha^2 \mu (1-\nu)} \ln \left(\frac{\beta \alpha \mu}{2\sigma} \right) \tag{11}$$

4.2 Sub-boundary energy

Consider nearly spherical subgrains with a diameter d' that occupy a deformed matrix (Fig. 2a). The number density of subgrains is

$$N = \frac{6}{\pi d'^3} \tag{12}$$

and the area of subgrain boundaries per unit volume is

$$A = N \cdot \pi d'^2 \cdot \frac{1}{2} = \frac{3}{d'} \tag{13}$$

The factor $1/2$ is included because the area of each subgrain wall is counted twice. The energy of sub-boundaries in a unit volume of the material can thus be written as

$$E_{sub} = \frac{3\gamma}{d'} \tag{14}$$

where γ is the sub-boundary energy per unit area.

The theory of dislocations gives

$$\gamma = \frac{\mu b^2}{4\pi(1-v)h} \xi(\eta) \tag{15}$$

$$\xi(\eta) \equiv \eta \coth \eta - \ln(2 \sinh \eta) \tag{16}$$

$$\eta \equiv \frac{\pi b}{\beta h} \tag{17}$$

where h is the mean dislocation spacing (Hirth & Lothe, 1982).

For a tilt boundary (Fig. 3), h and the misorientation angle θ are related by (Poirier, 1985)

$$\frac{b}{h} = 2 \tan \left(\frac{\theta}{2} \right) \simeq \theta \tag{18}$$

The last approximation is justified for low-angle boundaries. Then, Eq. (17) becomes

$$\eta = \frac{\pi}{\beta} \theta \ll 1 \tag{19}$$

Hence, the following approximations can be applied to Eq. (16):

$$\coth \eta \simeq \frac{1}{\eta}, \quad \sinh \eta \simeq \eta \tag{20}$$

Then, Eq. (15) becomes

$$\gamma = \frac{\lambda}{2} \mu b \theta \tag{21}$$

where

$$\lambda \equiv \frac{1}{2\pi(1-v)} \left[1 - \ln \left(\frac{2\pi\theta}{\beta} \right) \right] \tag{22}$$

The subgrain size is empirically expressed as (Takeuch & Argon, 1976; Twiss, 1977)

$$\frac{d'}{b} = K' \left(\frac{\sigma}{\mu} \right)^{-1} \tag{23}$$

where K' is a constant. A theoretical expression for K' is given below. Substituting Eqs. (21), (22), and (23) into Eq. (14), we have

$$E_{sub} = \frac{3\lambda\theta\sigma}{2K'} \tag{24}$$

4.3 Subgrain size

We consider a recovery process in which free dislocations with a dislocation density ρ rearrange into sub-boundaries. Conservation of the total dislocation length during subgrain formation requires

$$\rho = \frac{A}{h} \tag{25}$$

The right-hand side represents the length of dislocations in sub-boundaries. Using Eqs. (13) and (18), the above expression is modified to become

$$\rho = \frac{3\theta}{d'b} \tag{26}$$

Subgrains are formed if the total sub-boundary energy is smaller than the free dislocation energy (Twiss, 1977):

$$E_{disl} \geq E_{sub} \tag{27}$$

The equality represents the critical state for the initiation of subgrain formation. From Eqs. (4) and (14), this condition can be written as

$$\rho\zeta \geq \frac{3\gamma}{d'} \tag{28}$$

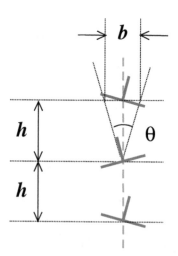

Fig. 3. Schematic illustration of a tilt boundary with a misorientation angle θ, dislocation spacing h, and Burgers vector b.

Substituting Eqs. (6), (7), and (21)–(22) into the above expression for ρ, ζ, and γ, respectively, Eq. (28) becomes

$$\left(\frac{\sigma}{\alpha\mu b}\right)^2 b \ln\left(\frac{\beta\alpha\mu}{2\sigma}\right) \geq \frac{3}{d'}\theta\left[1 - \ln\left(\frac{2\pi\theta}{\beta}\right)\right] \tag{29}$$

Equating Eqs. (26) and (6), we have

$$\left(\frac{\sigma}{\alpha\mu b}\right)^2 = \frac{3\theta}{d'b} \tag{30}$$

Then, Eq. (29) reduces to

$$\ln\left(\frac{\beta\alpha\mu}{2\sigma}\right) \geq \left[1 - \ln\left(\frac{2\pi\theta}{\beta}\right)\right] \tag{31}$$

The stability limit of θ is derived as

$$\theta \geq \frac{e}{\pi\alpha}\left(\frac{\sigma}{\mu}\right) \tag{32}$$

where e is the Napierian base. The equality gives the initial misorientation angle θ_i:

$$\theta_i = \frac{e}{\pi\alpha}\left(\frac{\sigma}{\mu}\right) \tag{33}$$

Applying θ_i to θ of Eq. (30), the initial subgrain size d'_i is obtained as

$$\frac{d'_i}{b} = \frac{3e\alpha}{\pi}\left(\frac{\sigma}{\mu}\right)^{-1} \tag{34}$$

Once the subgrain boundary is established, it functions as a dislocation sink because progressive subgrain misorientation is an energetically favorable process. We thus assume that the subgrain size is maintained during the subsequent misorientation. Substituting $d' = d'_i$ into Eq. (34), we obtain Eq. (23), where

$$K' = \frac{3e\alpha}{\pi} \tag{35}$$

Using Eqs. (22) and (35), the full expression of Eq. (24) is obtained as

$$E_{sub} = \frac{\pi\lambda\theta\sigma}{2e\alpha} = \frac{\theta}{4e\alpha(1-\nu)}\left[1 - \ln\left(\frac{2\pi\theta}{\beta}\right)\right]\sigma \tag{36}$$

4.4 Growth kinetics

The kinetic law of grain growth is generally written as

$$\dot{R} = MF \tag{37}$$

where M is the mobility of the grain boundary and F is the driving force. M depends on T as

$$M = \frac{bwD_{gb}}{kT} \tag{38}$$

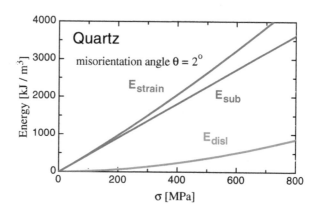

Fig. 4. Strain energy of quartz calculated using Eqs. (11) and (36). The physical parameters of quartz are given as (Shimizu, 2008) $\alpha = 3$ (Kohlstedt & Weathers, 1980), $\mu = 4.2 \times 10^4$ MPa, and $\nu = 0.15$ (Twiss, 1977). Because no data are available for β of quartz, we apply $\beta = 3$ of ionic crystals (Hirth & Lothe, 1982).

$$D_{gb} = D_{gb}^\circ \exp\left(-\frac{Q_{gb}}{RT}\right) \qquad (39)$$

where w is the boundary width, k is the Boltzmann constant, D_{gb} is the diffusion coefficient at the grain boundary, D_{gb}° is a constant, R is the gas constant, and Q_{gb} is the activation energy for grain boundary diffusion.

In a single-phase material, grain growth occurs to reduce the bulk strain energy and the energy of grain surfaces. Hence, Eq. (37) is written as

$$\dot{R} = M(F_{strain} + F_{surf}) \qquad (40)$$

where F_{strain} and F_{surf} represent the driving forces due to strain energy and surface energy (grain boundary energy), respectively. The strain energy in dynamically recrystallized materials is not homogeneous. The strain energy of deformed grains is given by the sum of E_{disl} in Eq. (11) and E_{sub} in Eq. (36), whereas newly recrystallized grains are almost strain free. This difference in strain energy drives grain growth. Hence,

$$F_{strain} = E_{strain} \qquad (41)$$

With increasing strain, free dislocations multiply and excess dislocations rearrange into sub-boundaries. Then, θ increases and the sub-boundary energy exceeds the free dislocation energy. Fig. 4 shows the calculations for quartz. When the average misorientation angle reaches several degrees, the following approximation can be used instead of Eq. (3):

$$E_{strain} \simeq E_{sub} \qquad (42)$$

5. Nucleation rate

In SGR nucleation, the nuclei are approximately the same size as the original subgrains. Thus, the number of potential nucleation sites per unit volume of crystals is given by Eq. (12) for intracrystalline nucleation and

$$N = \frac{6}{\pi d d'^2} \tag{43}$$

for nucleation at grain margins (Fig. 2b). The nucleation rate is scaled as

$$I = \frac{N}{\tau_c} \tag{44}$$

where τ_c is the interval of nucleation events.

The subgrain becomes a nucleus when the misorientation angle θ exceeds a critical value θ_c. The flux of dislocations that move toward the sub-boundary is given by ρu, where u is the climb velocity. The time required for dislocations to accumulate at the sub-boundary is equal to the nucleation cycle τ_c. From Eq. (18), a critical nucleus has a dislocation spacing of $h_c = b/\theta_c$; hence, the number of dislocations per unit area of the boundary is $1/h_c = \theta_c/b$. Dividing this value by the flux ρu, the nucleation cycle is evaluated as

$$\tau_c \simeq \frac{\theta_c}{b \rho u} \tag{45}$$

The climb velocity of dislocations is given by (Hirth & Lothe, 1982)

$$u = \frac{\sigma \Omega D_v}{lkT} \tag{46}$$

where Ω is the atomic volume, D_v is the self-diffusion coefficient, and l is a length scale given by

$$l \equiv \frac{b}{2\pi} \ln\left(\frac{r}{b}\right) \tag{47}$$

Using Eqs. (6) and (9), Eq. (47) can be rewritten as

$$l = \frac{b}{2\pi} \ln\left(\frac{\alpha \mu}{2\sigma}\right) \tag{48}$$

The temperature dependence of D_v is expressed as

$$D_v = D_v^\circ \exp\left(-\frac{Q_v}{RT}\right) \tag{49}$$

where D_v° is a constant and Q_v is the activation energy for volume diffusion.

Combining Eqs. (44)–(46), approximating Ω as b^3, and using Eqs. (12) and (23), we have

$$I = \frac{6}{\pi b K'^3 \alpha^2 \theta_c} \frac{\sigma D_v}{lkT} \left(\frac{\sigma}{\mu}\right)^5 \tag{50}$$

for intracrystalline nucleation. Using Eq. (43) instead of Eq. (12), the equation for marginal nucleation is obtained:

$$I = \frac{1}{d}\frac{6}{\pi K'^2 \alpha^2 \theta_c}\frac{\sigma D_v}{lkT}\left(\frac{\sigma}{\mu}\right)^4 \tag{51}$$

6. Scaling relation

Here, we neglect the surface energy term in Eq. (40) and assume

$$F = MF_{strain} \tag{52}$$

Combining Eq. (2) with Eqs. (52), (38), (39), (41), and (42), and using Eq. (36) and either Eq. (50) or Eq. (51), the steady-state grain size in continuous DRX is derived as

$$\frac{d}{b} = B\left(\frac{\sigma}{\mu}\right)^{-p}\left(\frac{wD_{gb}}{bD_v}\right)^{\frac{1}{m}} \tag{53}$$

where

$$p = \frac{5}{4} = 1.25, \; m = 4 \tag{54}$$

for intracrystalline nucleation and

$$p = \frac{4}{3} = 1.33, \; m = 3 \tag{55}$$

for marginal nucleation (Shimizu, 1998b; 2008). B is a non-dimensional constant given by

$$B = \left(\frac{a^4 \pi K'^{m-2}\alpha^2}{4}\frac{\lambda \theta \theta_c l}{b}\right)^{1/m} \tag{56}$$

Using Eq. (48), Eq. (56) can be rewritten as

$$B = \left[\frac{a^4 K'^{m-2}\alpha^2 \lambda \theta \theta_c}{8}\ln\left(\frac{\alpha\mu}{2\sigma}\right)\right]^{1/m} \tag{57}$$

Although σ is included in the right-hand side, the stress dependence of B is negligibly small. Using Eqs. (39) and (49), Eq. (53) can be re-expressed as

$$\frac{d}{b} = K°\left(\frac{\sigma}{\mu}\right)^{-p}\exp\left(-\frac{\Delta Q}{mRT}\right) \tag{58}$$

where

$$K° = B\left(\frac{wD°_{gb}}{bD°_v}\right)^{1/m} \tag{59}$$

and

$$\Delta Q = Q_{gb} - Q_c \tag{60}$$

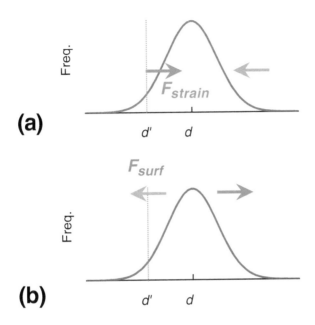

Fig. 5. Schematic representation of grain size evolution due to (a) strain-energy-driven grain growth and (b) surface-energy drag.

As Q_{gb} is generally smaller than Q_c, the recrystallized grain size is predicted to have a weak positive dependence on T. The constant K in Eq. (1) can now be written as a function of T:

$$K = K^{\circ} \exp\left(-\frac{\Delta Q}{mRT}\right) \tag{61}$$

7. Influence of surface energy

We now consider the influence of surface energy (grain boundary energy). In the case of surface-energy-driven grain coarsening in single-phase materials under static conditions (known as normal grain growth), large grains are energetically favorable and grow at the expense of small grains; the evolution of individual grain size has the opposite sense to that considered for DRX in Sec. 3 (Fig. 5). Therefore, when new grains grow by the strain-energy difference, the surface energy acts as a drag force.

In the theory of normal grain growth (Hillert, 1965), grain size evolution is described by

$$\dot{R}_k = Mc\Gamma\left(\frac{1}{R} - \frac{1}{R_k}\right) \tag{62}$$

where R_k and $\dot{R}_k$ are respectively the radius and the growth rate of the k-th grain and $c \sim 1$ is a statistical factor. If R_k is smaller (larger) than the mean radius R, the above expression becomes negative and the k-th grain shrinks (grows). By comparison with Eq. (37), the driving force for the growth of the k-th grain can be written as

$$2c\Gamma \left(\frac{1}{d} - \frac{1}{d_k} \right) \tag{63}$$

where d_k is the diameter of the k-th grain. In the nucleation and growth processes in DRX, the influence of the surface-energy drag is largest for small nuclei. Thus, we introduce a modified factor c' and express the surface-energy-driven force in Eq. (40) as

$$F_{surf} = 2c'\Gamma \left(\frac{1}{d} - \frac{1}{d'} \right) \simeq -2c'\Gamma \frac{1}{d'} \tag{64}$$

With this equation and Eq. (42), Eq. (40) can be approximated as

$$\dot{R} \simeq M \left(E_{sub} - \frac{2c'\Gamma}{d'} \right) \tag{65}$$

Using this equation, Eq. (56) can be modified as follows (the parameters p, m, and ΔQ remain the same).

$$B = \left[\frac{a^4 K'^{m-2}\alpha^2\theta_c}{8} \left(\frac{3\lambda\theta}{2} - \frac{2c'\Gamma}{b\mu} \right) \ln \left(\frac{\alpha\mu}{2\sigma} \right) \right]^{1/m} \tag{66}$$

8. Comparison of theory with experiments

8.1 Stress dependence of recrystallized grain size

In Fig. 6, p values of rock-forming minerals determined by triaxial or uniaxial or compression tests are plotted against the n-th power of dislocation creep flow laws ($\dot{\varepsilon} \propto \sigma^n$), which reflect the rate-controlling processes of dislocation creep; for climb-controlled creep, n is generally 3–5. The figure also shows the experimental result for a hexagonal Mg alloy (Magnox Al80), which was studied as a quartz analogue (De Bresser et al., 1998). The observed p values are almost independent of the power-law exponents and are well explained by the present model for continuous DRX.

8.2 Application to quartz

The theoretical model for the recrystallized grain size was applied to quartz using the equations presented in Sec. 6 (Shimizu, 2008; 2011). However, the previous model accounted only for strain energy; it neglected the effects of surface energy. Moreover, it turned out that the previous calculation involved a numerical error; when this error is corrected, the theoretical σ–d lines (Fig. 8 of Shimizu (2008)) shift to higher σ. Here, we recalculate the σ–d relation of quartz using the revised equations in Sec. 7.

Because experimentally deformed quartzite samples exhibit intracrystalline SGR at moderate stresses (Hirth & Tullis, 1992; Stipp & Tullis, 2003), we apply the intracrystalline nucleation model (Eq. 54). In addition to the material constants given in the caption of Fig. 4, we use $b = 5 \times 10^{-4}$ μm (Twiss, 1977), $\theta = 2°$, $\theta_c = 12°$, and D_v and D_{gb} of oxygen in β-quartz (Farver & Yund, 1991b; Giletti & Yund, 1984). For grain boundary energy, we use $\Gamma = 0.27$ Jm^{-2}

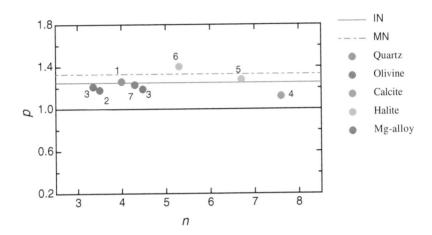

Fig. 6. Stress exponent p of recrystallized grain size plotted against the power-law exponent n of dislocation creep (IN: intracrystalline nucleation model; MN: marginal nucleation model). 1: Stipp & Tullis (2003) for p, Gleason & Tullis (1995) for n, 2: Karato et al. (1980) for p, Karato et al. (1986) for n, 3: van der Wal et al. (1993) for p, Chopra & Paterson (1984) for n, 4: Rutter (1995) for p, Schmid et al. (1980) for n, 5: Guillopé & Poirier (1979) for p and n, 6: Ter Heege et al. (2005) for p, Carter et al. (1993) for n, 7: De Bresser et al. (1998) for p and n. Microstructures of SGR and GBM are reported from all experiments except Ref. 3.

(Hiraga et al., 2007) and assume $c' = 1$. The steady-state grain size [μm] is then expressed a function of σ [MPa] and T [K] as

$$d = 1.82 \times 10^3 \times \sigma^{-1.25} \exp\left(\frac{7.25 \, \text{kJ/mol}}{RT}\right) \quad ; \beta-\text{quartz} \qquad (67)$$

In this expression, the weak stress dependence of B in Eq. (66) is neglected and $B = 1.01$ at $\sigma=50$ MPa is chosen as a representative value. The calculation results (Fig. 7a) agree well with the empirical data for β-quartz (Stipp & Tullis, 2003). For comparison, the σ–d relation based on the marginal nucleation model is also shown.

In Fig. 7(b), the theoretical model is extended to the α-quartz stability field in which D_v of oxygen in α-quartz (Farver & Yund, 1991a) is used and α- and β-quartz are assumed to have the same Q_v/Q_{gb} ratio. The recrystallized grain size of α-quartz is predicted to be

$$d = 9.98 \times 10^2 \times \sigma^{-1.25} \exp\left(\frac{12.4 \, \text{kJ/mol}}{RT}\right) \quad ; \alpha-\text{quartz} \qquad (68)$$

With decreasing temperature, the steady-state grain size shifts to higher stresses. If the empirical σ–d relation is directly applied to natural rocks that have deformed under low-T ($\leq 400°$C) metamorphic conditions, the stress states will be considerably underestimated.

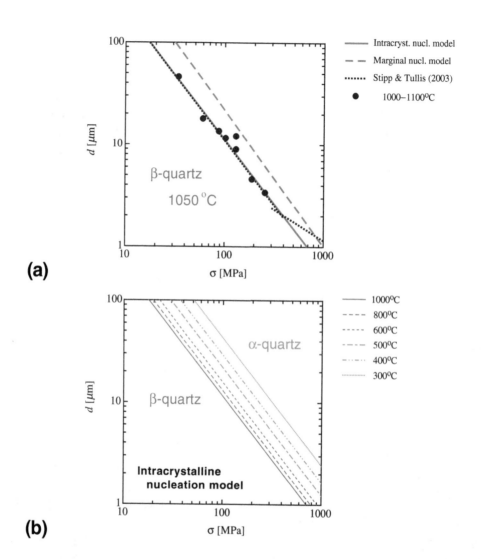

Fig. 7. Recrystallized grain size of quartz. (a) Theoretically calibrated σ–d relations for
β-quartz at 1050°C and the experimental results of Stipp & Tullis (2003). Solid line:
intracrystalline nucleation model. Dotted line: marginal nucleation model. Solid circles:
recrystallized grain size at 1000–1100°C after Stipp & Tullis (2003). Black dotted line:
empirical d–σ relation across the temperature range of 700–1100°C after Stipp & Tullis (2003).
(b) Theoretically predicted σ–d relations for β-quartz (blue lines, 1000–600°C) and α-quartz
(red lines, 500–300°C) using the intracrystalline nucleation model.

9. Summary

High-T dislocation creep of minerals is characterized by the occurrence of continuous DRX. The steady-state grain size is determined by the dynamic balance between SGR nucleation and grain growth by GBM. Surface energy acts as a drag force for strain-energy-driven GBM. The negative dependence of recrystallized grain size on stress is well explained by a theoretical model for continuous DRX. The theory also predicts a weak positive dependence of recrystallized grain size on temperature.

10. References

Carter, N.L.; Horseman, S.T.; Russel, J.E.; Handin, J. (1993). Rheology of rocksalt, *Journal of Structural Geology*, Vol. 15, 1257–1271.

Chopra P.N. & Paterson, M.S. (1984). The role of water in the deformation of dunite, *Journal of Geophysical Research* Vol. 89, 7861–7876.

De Bresser, J.H.P.; Peach, C.J.; Reijs, J.P.J.; Spiers, C.J. (1998). On dynamic recrystallization during solid state flow: effects of stress and temperature, *Geophysical Research Letters*, Vol. 25, 3457–3460.

Derby, B. (1992). Dynamic recrystallization: The steady state grain size, *Scripta Metallurgica and Materialia*, Vol. 27, 1581–1586.

Derby, B. & Ashby, M.F. (1987). On dynamic recrystallization, *Scripta Metallurgica*, Vol. 21, 879–884.

Farver, J. & Yund, R.A. (1991a). Oxygen diffusion in quartz: Dependence on temperature and water fugacity, *Chemical Geology*, Vol. 90, 55–70.

Farver, J. & Yund, R.A. (1991b). Measurement of oxygen grain boundary diffusion in natural, fine-grained, quartz aggregates, *Geochimica et Cosmochimica Acta*, Vol. 55, 1597–1607.

Gleason, G.C. & Tullis, J. (1995). A flow law for dislocation creep of quartz aggregates determined with the molten salt cell, *Tectonophysics*, Vol. 247, 1–23.

Giletti B.J. & Yund, R.A. (1984). Oxygen diffusion in quartz, *Journal of Geophysical Research*, Vol. 89, 4039–4046.

Guillopé, M. & Poirier, J.-P. (1979). Dynamic recrystallization during creep of single-crystalline halite: An experimental study, *Journal of Geophysical Research*, Vol. 84, 5557–5567.

Hillert, M. (1965). On the theory of normal and abnormal grain growth, *Acta Metallurgica*, Vol. 13, 227–238.

Hiraga, T.; Nishikawa, O.; Nagase, T.; Akizuki, M.; Kohlstedt, M. (2007). Interfacial energies for quartz and albite in pelitic schist, *Contributions to Mineralogy and Petrolgy*, Vol. 143, 663–672.

Hirth J.P. & Lothe J. (1982). *Theory of Dislocations*, Second edition, John Wiley & Sons, ISBN 0-471-09125-1, New York.

Hirth, G. & Tullis, J. (1992). Dislocation creep regimes in quartz aggregates, *Journal of Structural Geology*, Vol. 14, 145–159.

Humphreys, F.J. & Hatherly, M., (2004). *Recrystallization and Related Annealing Phenomena*, 2nd ed., Elsevier, ISBN 0-08-044164-5, Amsterdam.

Karato, S.; Paterson, M.S.; Fitz Gerald, J.D. (1986). Rheology of synthetic olivine aggregates: Influence of grain size and water, *Journal of Geophysical Research*, Vol. 91, 8151–8176.

Karato, S.; Toriumi, M.; Fujii, T. (1980). Dynamic recrystallization of olivine single crystals during high-temperature creep, *Geophysical Research Letters*, Vol. 7, 649–652.

Kohlstedt, D.L. & Weathers, M.S. (1980). Deformation-induced microstructures, paleopiezometers, and differential stress in deeply eroded fault zones, *Journal of Geophysical Research*, Vol. 85, 6269–6285.

Nabarro, F.R.N. (1987). *Theory of Crystal Dislocations*, Dover, ISBN 0-486-65488-5, New York.

Poirier, J.-P. (1985). *Creep of Crystals*, Cambridge University Press, ISBN 0-521-26177-5 (hardback) 0-521-27851 (paperback), Cambridge.

Rutter, E.H. (1995). Experimental study of the influence of stress, temperature, and strain on the dynamic recrystallization of Carrara marble, *Journal of Geophysical Research*, Vol. 100, 24651–24663.

Sakai, T. (1989). Dynamic recrystallization of metallic materials, In: *Rheology of solids and of the Earth*, Karato S. & Toriumi M. (Eds.), 284–307, Oxford University Press, ISBN 0-19-854497-9, Oxford.

Sakai, T. & Jonas, J.J. (1984). Dynamic recrystallization: mechanical and microstructural considerations, *Acta Metallurgica*, Vol. 32, 189–209.

Schmid, S.M.; Paterson, M.S.; Boland, J.N. (1980). High temperature flow and dynamic recrystallization in Carrara marble, *Tectonophysics*, Vol. 65, 245–280.

Stipp, M. & Tullis, J. (2003). The recrystallized grain size piezometer for quartz, *Geophysical Research Letters*, Vol. 30, 2088, doi:10.1029/2003GL018444.

Shimizu, I. (1998a). Lognormality of crystal size distribution in dynamic recrystallization, *FORMA*, Vol. 13, 1–11.

Shimizu, I. (1998b). Stress and temperature dependence of recrystallized grain size: A subgrain misorientation model, *Geophysical Research Letters*, Vol. 25, 4237–4240.

Shimizu, I. (1999). A stochastic model of grain size distribution during dynamic recrystallization, *Philosophical Magazine A*, Vol. 79, 1217–1231.

Shimizu, I. (2003). Grain size evolution in dynamic recrystallization. *Mater. Sci. Forum*, Vol. 426–432, Trans Tech Publ., Switzerland, 3587–3592.

Shimizu, I. (2008). Theories and applicability of grain size piezometers: The role of dynamic recrystallization mechanisms, *Journal of Structural Geology*, Vol. 30, 899–917.

Shimizu, I., (2011). Erratum to "Theories and applicability of grain size piezometers: The role of dynamic recrystallization mechanisms" [J Struct Geol 30 (2008) 899–917], *Journal of Structural Geology*, Vol. 33, 1136–1137.

Takeuchi, S. & Argon, A.S. (1976). Steady-state creep of single phase crystalline matter at high temperatures, *Journal of Materials Science*, Vol. 11, 1547–1555.

Ter Heege, J.H.; De Bresser, J.H.P.; Spiers, C.J. (2005a). Dynamic recrystallization of wet synthetic polycrystalline halite: dependence of grain size distribution on flow stress, temperature and strain, *Tectonophysics*, Vol. 396, 35–57.

Twiss, R.J. (1977). Theory and applicability of a recrystallized grain size paleopiezometer, *Pure and Applied Geophysics*, Vol. 115, 227–244.

van der Wal, D.; Chopra, P.; Drury, M.; Fitz Gerald, J. (1993). Relationships between dynamically recrystallized grain size and deformation conditions in experimentally deformed olivine rocks, *Geophysical Research Letters*, Vol. 20, 1479–1482.

3

Zircon Recrystallization History as a Function of the U-Content and Its Geochronologic Implications: Empirical Facts on Zircons from Romanian Carpathians and Dobrogea

Ioan Coriolan Balintoni and Constantin Balica
"Babeş-Bolyai" University, Cluj-Napoca
Romania

1. Introduction

According to Davies et al. (2003), "...the age of Earth and the time scale of pre-human events are central to a civilization's sense of origin and purpose. Therefore, the quest for precise and reliable geochronometers has had a scientific and cultural importance that few other enterprises can match". In this respect, since the beginning of the last century it has been recognized that long-lived radioactive decay systems provide the only valid means of quantifying geologic time.

One of the most reliable Earth's timekeeper has proven to be the mineral zircon, since it records the ages of Earth's earliest evolution stages, the oldest sediments, extinction episodes, mountain-building events and supercontinents' coalescence and dispersal (e.g. Rubatto and Hermann, 2007; Harley et al., 2007). Its widespread use in geochronology is based on the decay of uranium (U) and thorium (Th) to lead (Pb). They provide three distinct radioactive decay series involving the parent isotopes ^{238}U, ^{235}U and ^{232}Th and their daughter isotopes ^{206}Pb, ^{207}Pb and ^{208}Pb, respectively. Through the incorporation of U and Th at the time of growth, every zircon grain hosts three different clocks. In an ideal closed system, the three estimates would agree with each other within the analytical errors of measurements. However, in natural systems the zircon grains are not equally closed for Th and U with respect to post-crystallization effects. The usual approach in zircon geochronology is to consider the U-Pb system alone, as there is no natural non-nuclear ways of fractionating ^{235}U from ^{238}U. As the modern day-ratio of $^{235}U/^{238}U$ is well known (1/137.88) the need to actually analyze very low abundances of ^{235}U is obviated. Besides U and Th, zircon can incorporate some other incompatible elements such as P, Sc, Nb, Hf, Ti, and REE in trace (up to thousands of ppm) or minor (up to 3% wt) amounts. The primary control factor on the substitutions is the ionic radii of the substituting cations compared with Zr^{4+} and Si^{4+} cations. Substitutions that minimize strain effects on either or both sites will be favored. The crystal-chemical limitations are that Zr^{4+} in 8-fold coordination has an ionic radius of $84*10^{-3}$ nm and Si^{4+}, in tetrahedral coordination, has an ionic radius of $26*10^{-3}$ nm. U^{4+} (ionic radius of $10*10^{-2}$ nm in 8-fold coordination) and Th ($105*10^{-3}$ nm in 8-fold coordination) can be accommodated in the Zr^{4+} sites. Uranium concentrations are usually less than 5000 ppm and Th concentration less than 1000 ppm. Because of its ionic radius of $129*10^{-3}$ nm (8-fold coordination), Pb^{2+} is highly incompatible with growing zircon crystal lattice and therefore is not incorporated more than ppb levels, which is crucial in geochronology. Because of the same reason, Pb^{2+} can easily escape from zircon lattice when some conditions are fulfilled.

Based on idea of concordant ages between $^{235}U/^{207}Pb$ and $^{238}U/^{206}Pb$, Wetherill (1956) has developed the Concordia diagram. Quite soon though it has been shown (Tilton et al., 1957) that the concordance situations are rather rare and usually zircon shows evidence of discordance (disagreement between $^{235}U/^{237}Pb$ and $^{238}U/^{206}Pb$ ages) due to Pb loss caused by some post-crystallization geologic events.

Since then, the attempts made for understanding the causes of discordance had become the main preoccupation of the U-Pb geochronologists over the next quarter of century. Finally, it has been understood that discordance can be attributed to two major causes: (1) - mixing, in the analyzed sample volume, of discrete zones of different ages from within the same zircon grain; (2) - partial loss of radiogenic Pb by the entire zircon grain or by fractions of it. While the former cause can be bypassed by in situ dating, the second still hinders unequivocal results.

Continuously growing database of published in situ age data provided further insights into the behavior of U-Pb system in zircons: (1) - radiogenic Pb can be entirely lost and even the concordant data do not indicate the initial crystallization age of zircon; (2) – frequently, Pb loss is accompanied by Th loss; (3) – occasionally, even the U can be lost.

The disturbance of the isotope systematics in zircon is related to: (1) - amorphization; (2) - alteration; (3) - recrystallization.

The focus of the present contribution is on the discussion of the above three processes exemplified with their geochronological consequences on zircons from the Romanian Carpathians and Dobrogea. All data presented here were obtained by in situ dating through Laser Ablation-Inductively Coupled Plasma-Mass Specterometry (LA ICP-MS).

1.1 Amorphization process

This process is due to α-decay events associated with U and Th radioactive disintegration. Based on X-ray diffraction and High Resolution Transmited Electron Microscopy (HRTEM) analysis, Murakami et al. (1991) suggested three stages of damage accumulation in Sri Lankan zircon: in stage I (at $<3 \times 10^{18}$ α-decay events/g) the accumulation of isolated point defects predominate. These defects have the potential to recover through geologic time; stage II (at $3x10^{18}$ to $8x10^{18}$ α-decay events/g) is evidenced by crystalline regions with point defects and amorphous tracks caused by overlapped α-recoil nuclei; during stage III (at $> 8x10^{18}$ α-decay events/g) only aperiodic material can be screened by X-ray and electron diffraction.

Salje et al. (1999) suggested a two-phase transition during increasing amorphization process. The first phase is related to the percolation of amorphous material into the crystalline matrix and the second one to the percolation of crystalline material into the amorphous matrix.

Zircon Recrystallization History as a Function of the U-Content and Its Geochronologic Implications: Empirical
Facts on Zircons from Romanian Carpathians and Dobrogea

43

According to Nasdala et al. (2001) the α-decay events in the decay chains of U and Th cause the zircon amorphization. An α-particle generates about 120-130 Frenkel type defect pairs along penetration distances from 9.9 to 29.5μm. Recoils of heavy daughter nuclei are only a few hundred Å in length but the recoil damage clusters include 600-1200 Frenkel defect pairs. Spontaneous fission fragments produce heavy damage locally, but their contribution to the overall radiation damage is of minor importance because of their relative rarity. In the absence of recovery, radiation damage is stored in zircon, causing transformation from the crystalline to the metamict state. These authors propose the following stages of radiation damage accumulation.

1. Scattered nano-regions with high defect concentration. The amorphous component is still insignificant.
2. Moderately radiation damaged zircon in which amorphous nano-regions form a domain structure. Amorphous domains in a crystalline zircon or crystalline remnants in an amorphous matrix can be observed.
3. Entirely aperiodic zircon.

Ewing et al. (2003) speak about α-particles ionization processes over a range of 16 to 22 microns that produce several hundreds isolated atomic displacements. The associated α-recoils lose their energy during elastic collisions over 30 to 40 nm, producing localized collision cascades of 1000 to 2000 displacements. However, with increasing temperature the amorphization dose increases. This can be due either, by a decrease in the average cascade size caused by thermal relaxation or by a reduction in the surviving amorphous volume as a result of thermal recovery of irradiation induced defects.

According to Geisler et al. (2007), the amorphization of a crystalline zircon is characterized by transformation from an initial stage, where isolated amorphous domains are surrounded by slightly disordered crystalline material, to a more advanced stage of damage where few isolated, disordered, nano-crystalline islands occur in an amorphous matrix. These authors refer to Salje et al. (1999) when describing the crystalline- to-amorphous transformation as a geometrical phase transition. During the first phase transition, the amorphous domains form clusters that percolate over macroscopic length scales, whereas during the second transition, the crystalline domains cease to be interconnected. The first percolation point appears at an amorphous fraction of ~30% while the second percolation point appears when the amorphous fraction reaches ~ 70%. In other words, less than 30% amorphous fraction and less than 30% crystalline fraction can not be interconnected.

1.2 Alteration processes

According to Geisler et al. (2002), alteration is the interaction between the metamict zircon and fluids, including the meteoric ones (weathering processes), characterized by distinct chemical and structural changes. Geisler et al. (2003) describe two anomalous stages in the alteration rate with increasing degree of amorphization. The first stage takes place when the amorphous domains form interconnected clusters within the zircon structure, namely at the first percolation point suggested by Salje et al. (1999). At this point, the percolation interfaces that represent low-density areas between crystalline and amorphous domains open high diffusivity pathways. A new dramatic increase in alteration rate is observed at the second percolation point of Salje et al. (1999). Around this point, a network of nanometer-size regions of depleted matter interconnects discrete amorphous domains without crystalline dams.

According to Geisler et al. (2007), the structural changes that take place in a zircon grain are defined by the fact that above the 200° C limit the thermal recrystallization front generated by epitaxial reordering moves inward the zircon crystal. Bellow the 200° C threshold, the thermal recrystallization involves enhanced defect diffusion only along the reordering front. The above model hypothesizes an increase in the effective diffusion of any species within the zircon lattice with increasing α-decay dose or of amorphization degree.

1.3 Recrystallization

There are several types of recrystallization processes that take place in zircons and several different opinions in terms of its significance exist on this matter.

According to Nasdala et al. (2001), *recrystallization* is a "re-growth" process in the crystallographic sense. During recrystallization a new zircon lattice forms along a crystallization front, typically replacing a more disordered and polluted zircon. Recrystallization leads to healing of the radiation damage and partial or complete resetting of the U-Pb isotopic system. According to the same authors, *annealing* requires only re-formation of disrupted bonds by re-ordering of nearest neighboring atoms, thus annealing of zircon structure is not necessarily associated with any Pb loss. As defined in the literature, *recrystallization* presumes an epitaxial migration of an interface between an ordered region in zircon and a metamict vicinity, therefore it could be described as a defect annihilation or point defect diffusion process. In a general view, Nasdala et al. (2001) enumerate the following thermal recovery mechanisms: (1) - point defect diffusion in the crystalline and amorphous phase; (2) - epitaxial growth of crystalline residuals; (3) - random nucleation in the amorphous phase.

Ewing et al. (2003) put forward a more comprehensive description of the concept of recrystallization, while offering the foundation for a clear distinction between *Type I* and *TypeII* recrystallization. Type I recrystallization is purely thermal and occurs on time scales longer than cascade quench time. This behavior is due to point defect diffusion and epitaxial migration of the crystalline residuals toward the amorphous domains. It can be seen that Type I recrystallization, as described by these authors, covers both the recrystallization and annealing processes of Nasdala et al. (2001). Type I recrystallization prevails over longer periods of time or at higher temperatures and becomes particularly important in natural specimens stored at ambient conditions for geological periods. In moderately damaged zircons, two stages of recovery process have been described. The first stage is defined by the recovery of short-range order and point defect recombination. It occurs bellow ~ 600° C (Farges, 1994) or ~727° C (Geisler et al., 2001). The second stage occurs at higher temperatures and is caused by epitaxial recrystallization along the internal crystalline-amorphous boundaries. It is worth noting that an initially moderately damaged zircon grain consists of distorted crystalline phases embedded in an aperiodic matrix at the completion of the first phase. In contrast to the moderately damaged specimens, heavily damaged or amorphous zircon segregates into its constituent oxides at higher temperatures and recrystallization takes place in three stages: (1) - decomposition of amorphous zircon into tetragonal ZrO_2 crystallites and amorphous SiO_2; (2) - tetragonal to monoclinic phase

Zircon Recrystallization History as a Function of the U-Content and Its Geochronologic Implications: Empirical
Facts on Zircons from Romanian Carpathians and Dobrogea

45

transformation in the ZrO_2 crystallites; (3) - formation of coarse-grained (several hundred µm) randomly oriented, polycrystalline zircon. This process is highly improbable to take place in natural zircon and it excludes the idea of new zircon nucleation within amorphous phase as postulated by Nasdala et al. (2001).

Type II recrystallization occurs as a nearly instantaneous process during irradiation and can be divided into two distinct phases. The first, Type *IIa* recrystallization, is due to increased mobility of interstitials and other point defects during irradiation. The irradiation enhanced diffusion leads to a greater degree of point defect recombination and annihilation. Point defect annihilation is most effective at structural boundaries between amorphous and damaged, but still crystalline regions. Type *IIb* recrystallization occurs within individual displacement cascades. Disordered and highly energetic material can epitaxially recrystallize at the cascade peripheries along with the cooling of the displacement cascade to the ambient temperature. During irradiation, both *Type I* and *Type II* recrystallization processes contribute to the dynamic recovery of zircon, but some mechanisms can prevail over others in certain temperature regimes. Obviously, at the Earth's surface conditions, both Type I and Type II recrystallization processes are less effective than the irradiation damage over an α-dose range, allowing amorphization accumulation in time.

Geisler et al. (2007) proposed two more kinds of zircon recrystallization, fluid or melt assisted. The Hf, U and Th content of zircon can be also distributed in solid solutions between zircon and hafnon ($HfSiO_4$), coffinite ($USiO_4$), thorite ($ThSiO_4$). Because solid-state exsolution structures have not been yet reported in zircon, it is argued that such solid solutions are metastable after cooling and characterized by structural strain. Such structural strain enhances surface reactivity and thus the dissolution rate. We note that the effects of the structural strain are supplementary added to irradiation damage, increasing the reactivity between the zircon and fluids. It was shown (Geisler et al., 2003) that the treatment of radiation-damaged zircon crystals in various aqueous solutions produces inward-penetrating, irregular, and curved reaction domains that resemble those found in natural zircon. Recrystallization of zircon on the expense of the amorphous phase inside the reacted domains occur at experimental temperatures above 200° C. Recrystallization of amorphous zircon dramatically reduces the molar volume of the reacted areas inducing a strain that is partially released by fracturing. Porous structure at nanometer-scale level is also very likely to occur. The nanoporosity created between the crystallites provides pathways for chemical exchange between reaction front and the fluids. This is the diffusion-reaction process in which a moving recrystallization front follows at some distance behind the percolation-controlled, inward diffusion of a hydrous species. According to Geisler et al. (2007), a prerequisite for the diffusion-reaction process is the presence of more than 30 % amorphous fraction, which is of an interconnected amorphous network.

The coupled dissolution-reprecipitation is a process by which the breaking of the bonds and dissolution is accompanied by contemporaneous nucleation and precipitation of new zircon. The coupled dissolution-reprecipitation process is independent of the absolute solubility of zircon in natural aqueous fluids, which is very low (Tromans, 2006). This can result in a complete replacement of one zircon crystal by a new one within the same space, without losing the external shape or crystal morphology (Putnis, 2002; Putnis et al., 2005). The chemical exchange between the dissolution-reprecipitation front and external fluid is maintained by the formation of porosity. This porosity results from the higher solubility and the higher molar volume of the dissolved parent zircon, as compared with the more pure zircon, chemically reprecipitated . Since the dissolution of a metastable zircon solid solution is kinetically favored by structural strain, the radiation damage may also enhance kinetically a coupled dissolution-reprecipitation process. Considering the two proposed mechanisms, the reaction of zircon with fluids and melts provides an effective way of its re-equilibration.

1.4 Temperature conditions of zircon recrystallization

According to Mezger and Krogstad (1997), the 600-650° C temperature interval is a good estimate for recrystallization of damaged zircon. Consequently, lattice damage through α-decay and spontaneous fission may accumulate bellow this temperature. The α-recoil tracks in minerals recrystallize at similar temperature as tracks formed by spontaneous fission fragments (e. g. Murakami et al., 1991). And because fission tracks are retained in zircon up to 200-250° C (Tagami and Shimada, 1996), the α-events damage can also be recovered immediately over 200-250° C (Nasdala et al., 2001). According to Ewing et al. (2003), "temperatures as low as 100-200° C seem to be sufficient to produce measurable recovery over extremely long time scales". Such statements are valid knowing that the quantity of radioactive elements is continuously decreasing in zircon through fission process. The same authors suggest that the critical amorphization temperature for zircon should have an upper limit of 460 K, or 187° C. This temperature is close to the interval indicated by Tagami and Shimada (1996) for recovery initiation. However, Ewing et al. (2003) say that "depending on the mass of the incident ions, the critical amorphization temperature for zircon is between 527 and 750° C". This statement can be understood in the context of external ionic bombardment. Commenting the data of Meldrum et al. (1998), Cherniak and Watson (2003) say that "the critical amorphization temperature for zircons with 1,000 ppm U is about 360° C, and only about 20° C higher for zircons with as much as 10,000 ppm U; and it varies as a function of zircon age by less than a degree per billion years for a given U content. Zircon exposed to temperatures bellow the critical amorphization temperature can accumulate radiation damage, but only over long time scales".

1.5 Isotope systems resetting

According to Mezger and Krogstad (1997) Pb-loss in zircon may occur in four distinct ways:(1)-diffusion in metamict zircon;(2)-diffusion in pristine zircon;(3)-leaching from metamict zircon;(4)-recrystallization of metamict zircon. Pb-loss can be also accompanied by U and Th loss. As mentioned, Geisler et al. (2007) discussing re-equilibration of zircon in aqueous fluids and melts describe two more processes in such environments: (5) - diffusion-reaction process and (6) - coupled dissolution-re-precipitation process. Possibly, leaching from metamict zircon is always accompanied by diffusion-reaction.

According to Cherniak and Watson (2003), cation diffusion in pristine zircon appears to be exceedingly slow under normal crustal conditions. They mention that the closure temperature for Pb in zircon of 100 μm effective diffusion radius for a cooling rate of 10° C/Ma is 991° C. Field based studies showed that Pb-diffusion in the pristine zircon lattice is insignificant around 950-1000° C (e.g. Black et al., 1986; Williams, 1992). Thus, we do not further explore the possibility of isotope systems resetting in pristine zircon.

Zircon Recrystallization History as a Function of the U-Content and Its Geochronologic Implications: Empirical
Facts on Zircons from Romanian Carpathians and Dobrogea

47

Regarding the diffusion in metamict zircon, Davies and Paces (1990) and Heaman et al. (1992) observed that even metamict zircon can retain Pb if the temperature is low enough to inhibit recrystallization or there has been no chemical attack of the metamict parts.In conclusion, resetting of U, Th, and Pb istope systems is strongly dependent on leaching and recrystallization of metamict zircon.

Accumulation of radiation damage in zircon is a competition between the α-dose induced disorder (as a function of its content in U and Th) and recrystallization processes. With increasing temperature the irradiation defects become gradually neutralized by instantaneous reordering of lattice, which is equivalent with *Type II* recrystallization. When no new radiation defects accumulate, the critical amorphization temperature is reached. However, we should stress out that most of the previously amorphizated lattice volume is preserved at the critical amorphization temperature. As mentioned before, there is little consensus with respect to the critical amorphization temperature, which is also a function of the α-dose. If the data of Meldrum et al. (1998) are applicable for natural zircon, then the accumulation of amorphization in grains containing 10,000 ppm U is possible bellow 380° C only. In addition to that, if zircon contains 100 ppm U or less, the amorphization becomes improbable (e.g. Mezger and Krogstad, 1997), as a consequence of the low α-dose. Above the critical amorphization temperature and especially above 600-650° C, *Type I* recrystallization will recover the zircon's structure (e.g. Mezger and Krogstad, 1997). During this process, Pb, Th, and U will be variably lost from zircon lattice. If the metamict zircon interacts with fluids, Pb, Th, and U can be lost at temperatures lower than 200° C due to structural recovery front of low temperature *Type I* recrystallization, and over 200° C due to a moving recovery front of high temperature *Type I* recrystallization (Geisler et al., 2007). The diffusion-reaction process causes only partial loss of radiogenic Pb. According to Geisler et al. (2007) an unambiguous chemical indication of the alteration of radiation-damaged zircon by a diffusion-reaction process is the enrichment in Ca, Al, and Fe, and also in common Pb. Enrichment in common Pb can be easily recognized in the recent geochronologic data sets. Briefly, to arrive to a more or less metamict state, which is a function of its U and Th content, and to allow to its isotope systematics to be disturbed, zircon should have: (1) - remained, in geological time terms, bellow the critical amorphization temperature; (2) –reheated subsequently over the critical amorphization temperature; (3) – interacted with aqueous fluids independently of temperature and/or with melts.

2. U, Th, and Pb isotope systems in granites

Dating granites is not always an easy enterprise. They can have a lot of inherited zircons, some of these could have recrystallized in granitic magma and the true magmatic zircons could have been isotopically destabilized at various degrees, depending on their U content and/or on the interaction with fluids. Different aspects of these questions will be exemplified with the Carpathian granites and migmatites.

1. *Variscan granites in the Danubian Domain of Romanian South Carpathians.* Peri-Amazonian basement of the Alpine Danubian Domain of South Carpathians (Balintoni et al., 2011) was massively intruded by granite bodies and their accompanying dyke-swarms (Berza and Seghedi, 1983). There are Cadomian granites with their migmatic escort (Grünenfelder et al., 1983; Liégeois et al., 1996; Balintoni et al., 2011) and Variscan bodies accompanied by cross-cutting dykes (Balica et al., 2007; Balintoni et al., 2011).

From the Variscan intrusions we will further consider the *Buta* pluton (sample 266)[*] for its low U and isotopically undisturbed zircons in comparison with *Cherbelezu* and *Sfârdinu* plutons defined by high U and strongly isotopically disturbed zircons.

The grains in sample 266 are characterized by: (1) - an U-content in zircon dominantly less than 400 ppm; (2) - a high $^{206}Pb/^{204}Pb$ ratio; (3) - a small U/Th ratio; (4) - nearly concordant ages (Fig. 1).

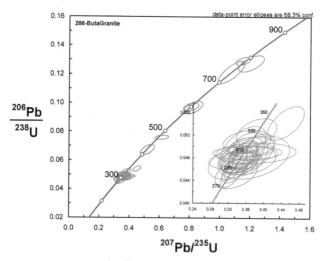

Fig. 1. Concordia projection for sample 266. Good concordances and great concentration around 300 Ma can be noticed. Inset: distribution of ages around 300 Ma.

The vast majority of these zircons represent originally Variscan magmatic grains. Twenty seven $^{206}Pb/^{238}U$ apparent ages ranging between 288.9±7.7 Ma and 314.8±4.3 Ma yielded a crystallization weighted mean age of 303.7±2.4 Ma and a Concordia age of 303.8±0.85 Ma (Fig. 2a, b).

Cherbelezu puton data presented further on were yielded by sample 227. The grains from sample 227 are characterized by two data sets with different parameters: older than 315.2±3.3 Ma and younger than this age. The younger grains with ages ranging between 295.2±3.1 Ma and 122.5±15.1 Ma show: (1) - higher U-content; (2) - higher content in ^{204}Pb; (3) - much greater discordances than the older ones (Table 1 and Fig. 3).

In our interpretation, the older ages suggest inherited recrystallized grains in the Variscan granitic magma. Their isotope systems were completely reset by high temperature *Type I* recrystallization without fluid intervention. The zircons lost all their radiogenic Pb (i.e., good concordance of the data) and partially Th. The available data doesn't allow us to draw some conclusion with respect to any potential U loss. The younger grains most likely represent true Variscan magmatic zircons. Due to their high U content they were amorphizated post 300 Ma and partially lost radiogenic Pb (i.e., poor concordance of the data) and Th during the Alpine thermotectonic events, while gaining ^{204}Pb. The isotope

[*] Analytical data are available upon request from the authors

Zircon Recrystallization History as a Function of the U-Content and Its Geochronologic Implications: Empirical Facts on Zircons from Romanian Carpathians and Dobrogea

49

systems were probably disturbed in the presence of fluids and amorphization grade reached the first percolation point of Salje et al. (1999). The data suggest that a diffusion-reaction process was active due to structural recovery front of low temperature *Type I* recrystallization, bellow 200° C. The true zircons' crystallization time was probably less than 315.2±3.3 Ma and more than 295.2±3.1 Ma. Thus we suggest for Cherbelezu pluton a crystallization age closed to that of Buta pluton.

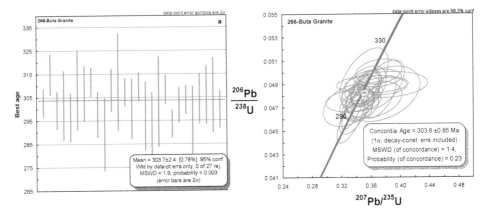

Fig. 2a, b. Weighted mean age and Concordia age for sample 266.

	Older grains	Younger grains
U average content (ppm)	3990	13623
204Pb average content (ppm)	3.46	87.9
U/Th average ratio	8.5	6.3

Table 1. Comparative isotope parameters between the two sets of grains in sample 227.

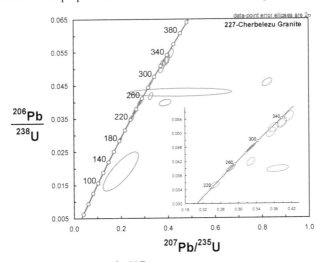

Fig. 3. Concordia projection for sample 227.

The data presented for *Sfârdinu* pluton were provided by sample 229. The grains pertaining to this sample are characterized by: (1) - high U-content (4413 ppm average); (2) - high [204]Pb-content; (3) - great discordances for most of the ages, except for 2 out of 50 (Fig. 4); (4) - variable U/Th ratios, generally over the normal ratios in undisturbed magmatic zircon.

Considering the above observations, we infer that all the analyzed crystals represent inherited zircons disturbed initially during the Variscan orogeny, followed by the Alpine thermotectonic events. In their present state all the grains show isotopic disturbance assisted by fluids. Amorphization grade reached the first percolation point of Salje et al. (1999) and grains probably remained bellow 200° C along their entire post Variscan history. Clearly, also in this case, the grains with more than 5000 ppm U gained much more [204]Pb than the grains with less than 5000 ppm U. The crystallization age of Sfârdinu pluton is difficult to ascertain. If we consider the most concordant age sets (between 305.4 and 318.6 and between 292.1 and 305.6 Ma) we get a weighted mean age of 301.5±6 Ma (Fig.5).

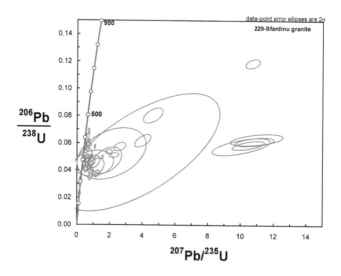

Fig. 4. Concordia projection for sample 229.

Zircon Recrystallization History as a Function of the U-Content and Its Geochronologic Implications: Empirical
Facts on Zircons from Romanian Carpathians and Dobrogea

51

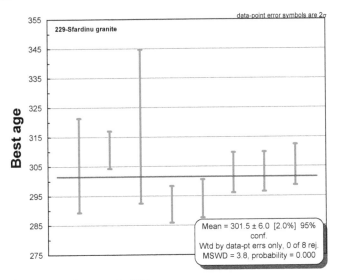

Fig. 5. Weighted mean age for sample 229.

2. *Cadomian granites and migmatites.* Grünenfelder et al. (1983), Liégeois et al. (1996), and Balintoni et al. (2011) dated the Tismana, Şuşiţa, Novaci, and Olteţ plutons and their results bracketed the ages around 600 Ma. These plutons cross-cut a dense swarm of migmatic dykes characterized by black K-feldspar grains (Berza and Seghedi, 1983). Out of the four Cadomian plutons we exemplify the relationship between U-content and isotope disturbance in Şuşiţa pluton. Two samples (277A and 277B) collected several km apart each other show quite different isotope data.

Using the data from sample 227A, Balintoni et al. (2011) obtained a crystallization weighted mean age of 591.0±3.5 Ma and a Concordia age of 591.6±1.8 Ma (Fig. 6a, b).

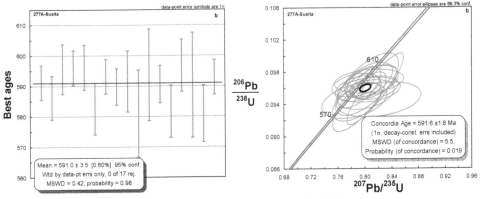

Fig. 6a, b. Weighted mean age and Concordia age for sample 277A.

In comparison with the grains from sample 277A, the grains from sample 277B show: (1) - higher U-content; (2) - higher U/Th ratio; (3) - lower $^{206}Pb/^{204}Pb$ ratio; (4) - greater discordances (Fig.7); (5) - complete lack of protolith ages. Judging from the age discordances, is safe to assume that all the ages were partially reset. The comparative parameters are presented in Table 2.

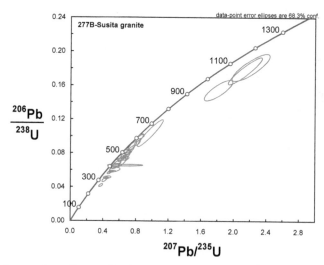

Fig. 7. Spread of ages along Concordia for sample 277B. No ages around 600 Ma.

	Sample 277A	Sample 277B
U average content (ppm)	392	1412
U/Th average ratio	3.2	22.7
$^{206}Pb/^{204}Pb$ average ratio	71374	42429

Table 2. Comparative isotope parameters between the grains in samples 277A and 277B.

Zircons from sample 277B lost radiogenic Pb and Th, and gained ^{204}Pb. There is a direct corellation between U-content, ^{204}Pb gain, age rejuvenation, and concordance deterioration. Therefore, for ages bellow 400 Ma the U average content is of 1883 ppm and $^{206}Pb/^{204}Pb$ average ratio is 27351. For ages between 400 and 500 Ma, the average content of U is of 1235 ppm and the $^{206}Pb/^{204}Pb$ average ratio is 36560. For sample 277B we infer an amorphization grade of up to 40 % and *Type I* recrystallization by point defect diffusion under moderate influence of fluids, as the main recrystallization mechanism. According to Geisler et al. (2007), we assume a diffusion-reaction process at temperatures bellow 200° C. The inferences made for the sample 277B are strongly sustained by the data from the samples 334 (Fig. 8) and 10-406 (Fig. 9) representing Cadomian migmatic material.

The main conclusion that can be drawn is that by increasing the U-content, the isotopic parameters become gradually deteriorated (e.g., discordances in Figs. 8 and 9), while making a distinction between the Variscan and Alpine reworking virtually impossible. At higher U contents the radiogenic Pb and Th are lost in greater quantities and concomitantly more common Pb is added as portrayed in Table 3.

Zircon Recrystallization History as a Function of the U-Content and Its Geochronologic Implications: Empirical Facts on Zircons from Romanian Carpathians and Dobrogea

53

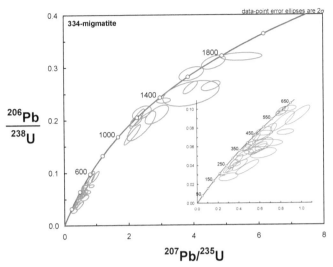

Fig. 8. Spread of ages along Concordia for sample 334. All the magamatic ages have been reset.

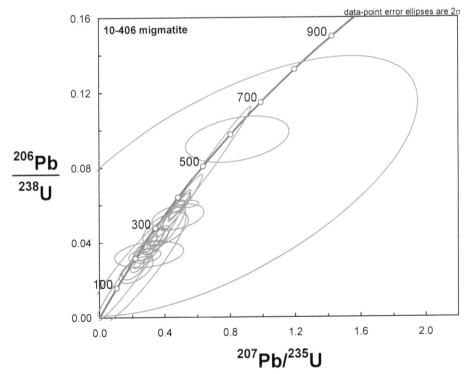

Fig. 9. Spread of ages along Concordia for sample 10-406. Again all the magmatic ages have been reset.

Sample 334		
Ages (Ma)	U-average content (ppm)	U/Th average ratio
>617	385	2.6
617-383	1236	23.9
<383	2049	55.5
Sample 10-406	4465	115.7

Table 3. Comparative isotope parameters between sets of grains from sample 334 and the same parameters for the grains from sample 10-406.

No age was completely reset and the involvement of fluids in these processes is highly suspected. We postulate a poor zircon lattice recovery by defect diffusion processes but the amorphizated material did not recrystallized. During post Cadomian events the grains remained bellow the amorphization critical temperature, very likely bellow 200° C. In the case of sample 10-406 the massive loss of Th and gain of common Pb can be interpreted as an amorphization grade above the second percolation point of Salje et al. (1999), that is more than 70 % amorphization.

3. U, Th, and Pb isotope systems in orthogneisses, metamorphosed under medium grade conditions

The igneous protoliths of orthogneisses are isotopically stabilized during metamorphism. However, as a function of the U-content, temperature history, and fluids intervention, the isotope systems can potentially be reset. Several examples from Apuseni Mountains and East Carpathians will illustrate the behavior of zircon in orthogneisses during Ordovician initial metamorphism and during later Variscan thermotectonic events.

1. *Orthogneisses from the basement of Someş pre-Alpine terrane Apuseni Mountains.* The geochronology of the Apuseni Mountains pre-Alpine terranes has been detailed by Balintoni et al. (2010b). The Someş terrane basement of Ordovician age was intruded by Variscan granites when it functioned as an upper plate. We will further discuss the data from the samples 166 and 167, representing the same orthogneiss body.

The grains from the sample 166 show good concordances (Fig.10), a low ^{204}Pb, small U/Th ratios, and variable U-content. Only three grains show strong reset of U/Pb isotopic system during the Variscan thermotectonic events, with an U average-content of 1642 ppm. The majority of zircon grains preserve Ordovician ages, yielding a weighted mean age of 452.3±5.2 Ma and a Concordia age of 452.4± 3.7 Ma (Fig. 11a, b). The data from sample 166 suggest a Type I recrystallization process of higher temperature of the amorphizated grains, without fluids intervention.

At an U average-content of 2002 ppm, all the grains from the sample 167 lost the radiogenic Pb while the youngest of them gained ^{204}Pb. Rejuvenation lead to the deterioration of the concordances (Fig. 12) but the U/Th ratio remained relatively constant. This is further proof that amorphization grade and resetting of the isotopic systems are a function of the U-content. In the case of sample 167, we advocate for a *Type I* recrystallization at a temperature over the critical amorphization threshold.

Zircon Recrystallization History as a Function of the U-Content and Its Geochronologic Implications: Empirical
Facts on Zircons from Romanian Carpathians and Dobrogea

55

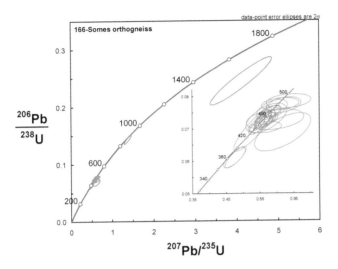

Fig. 10. Concordia projection for sample 166.Concordant ages around 450 Ma.

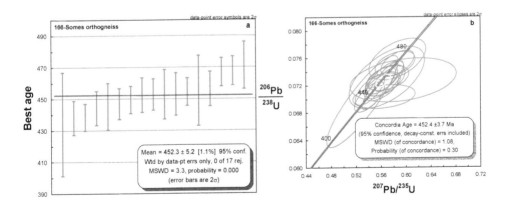

Fig. 11a, b. Weighted mean age and Concordia age for sample 166.

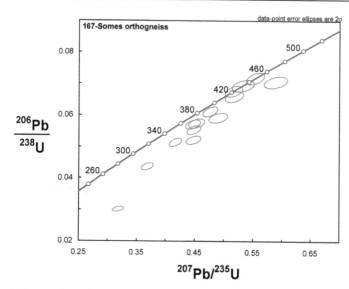

Fig. 12. Spread of ages along Concordia for sample 167.Reset and discordant ages at higher U-content.

2. *Orthogneisses from the basement of the East Carpathians pre-Alpine terranes.* Preliminary geochronologic data on the basement of the pre-Alpine terranes from East Carpathians were presented by Balintoni et al. (2009). In this contribution we will focus mainly on the data from the basement of the Tulgheş terrane and from the basement of the Rebra terrane represented by Negrişoara metramorphic unit. These terranes composed the median part of the Variscan nappe pile and were affected by retrogression down to the chlorite grade temperature (Balintoni, 1997) when the K/Ar ages were reset around 300 Ma (Kräutner et al., 1976).

Tulgheş orthogneiss, sample 10-476. Except a single anomalous age, the data in sample 10-476 show: (1) - good concordances (Fig 13); (2) - low U/Th ratio; (3) - high $^{206}Pb/^{204}Pb$ ratio; (4) - low U-content; (5) - no age younger than the Ordovician. Considering all the Ordovician ages, these can be divided in two distinct groups: (i) between 449.6 and 469.0 Ma; (ii) between 474.4 and 488.8Ma. The younger set corresponds to an U average - content of 203 ppm and to an U/Th average ratio of 3.7., while the older set corresponds to an U average-content of 185 ppm and to an U/Th average ratio of 3.0. The above data suggest a slight increase in the U-content and a decrease in the Th content toward younger ages. Because there is little evidence for any disturbance in the isotopic systems we interpret the age range as an evolution of the magmatic system from its source to the crystallization time. The two data sets yielded a younger $^{206}Pb/^{238}U$ weighted mean age of 462.6±3.1 Ma and an older one of 478.3±5.5 Ma (Fig. 14a, b). We consider the first age to be a better candidate for the protolith crystallization age. There is no doubt that the dated grains constituted isotopically closed systems post protolith crystallization. Therefore, we conclude that bellow the concentration of 250 ppm U, the zircon lattice is prone to recovering even at low temperatures, by *Type* II recrystallization, while the radiation damage and the amorphization can not accumulate.

Zircon Recrystallization History as a Function of the U-Content and Its Geochronologic Implications: Empirical
Facts on Zircons from Romanian Carpathians and Dobrogea

57

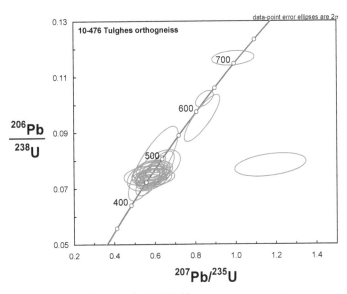

Fig. 13. Concordia projection for sample 10-476.No one age was reset.

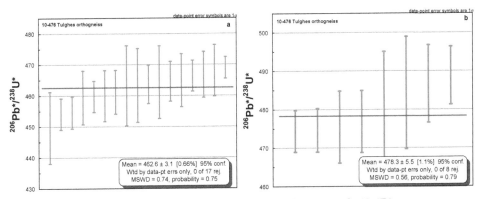

Fig. 14a, b. Weighted mean ages for the two sets of ages from sample 10-476.

Pietrosu Bistriței orthogneiss, Negrişoara metamorphic unit, sample 10-475.

The data from this sample are similar to data from sample 10-476.The younger grains show 226 ppm U in average and an U/Th average ratio of 8.4, while the older grains have an average concentration of 226 ppm U and an U/Th average ratio of 5.5. The younger data set yielded a ^{206}Pb/^{238}U weighted mean age of 461.5±5.2 Ma and the older one yielded 477.8±4.2 Ma (Fig. 15a, b). We interpret these ages identically to ages from sample 10-476. As in the case of the previous sample, there is no evidence for zircon damage due to radiation at an U-content bellow 300 ppm.

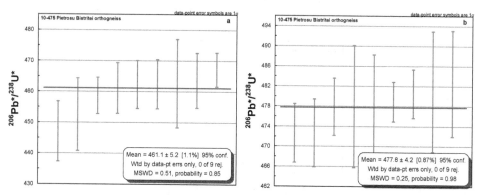

Fig. 15a, b. Weighted mean ages for the two sets of ages from sample 10-475.

4. U, Th and Pb isotope systems in orthogneiss zircons affected by Variscan eclogite-facies metamorphism

Medaris et al. (2003) argued for a Variscan age of the eclogite-grade metamorphism known from the basement of the Sebeş-Lotru pre-Alpine terrane (Iancu et al., 1998; Săbău and Massonne, 2003). Balintoni et al. (2010c) published geochronologic data that revealed the composite nature of the Sebeş-Lotru terrane basement, which consists of Cadomian and Ordovician (Caledonian) igneous protoliths. In the following paragraphs the Cadomian *Frumosu* orthogneiss and *Tău* Ordovician orthogneiss will be discussed.

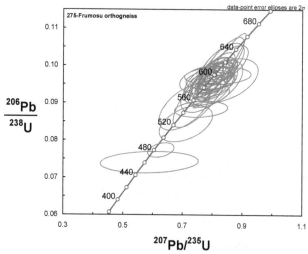

Fig. 16. Concordia projection for sample 275

Frumosu orthogneiss .The grains from sample 275 are characterized by: (1) - very good concordances (Fig 16); (2) - low U/Th ratios; (3) - high $^{206}Pb/^{204}Pb$ ratios; (4) - low U-contents. The U average-content for all grains is around 300 ppm. From 36 dated zircon grains, only 5 of them show Pb loss and 3 of them exhibit Th loss in various degrees. The ages tend to cluster on

Zircon Recrystallization History as a Function of the U-Content and Its Geochronologic Implications: Empirical
Facts on Zircons from Romanian Carpathians and Dobrogea

59

Concordia diagram and can be divided in two sets as in East Carpathians orthogneisses, between 568 and 593 Ma and between 597 and 618 Ma. The first data set yielded a $^{206}Pb/^{238}U$ weighted mean age of 584.8±3.6 Ma and the second set yielded a mean age of 606.6±4.4 Ma (Fig. 17a, b). We consider the younger age closer to the crystallization time of igneous protolith while the slightly older mean age is interpreted as an early crystallization event during melt genesis. The main conclusion is that even under the eclogite metamorphic facies conditions, the zircons with less than 300 ppm U do not show isotope systems resetting.

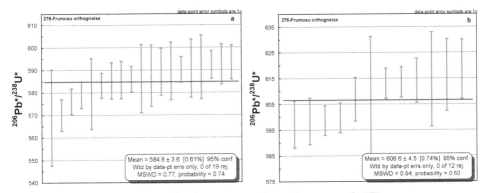

Fig. 17a, b. Weighted mean ages for the two sets of ages from sample 275.

Tău orthogneiss crops out in the median part of the Sebeş valley as a component of the Cumpăna Ordovician metamorfic unit.

The grains from sample 272 are characterized by: (1) - over 1000 ppm U in all the grains; (2) - a deterioration of concordances, yet acceptable for many grains (Fig.18); (3) - a single grain from 36 preserving a protolith age; (4) - Th loss toward the younger ages; (5) - ^{204}Pb gain is in variable quantities and not recorded by all grains.

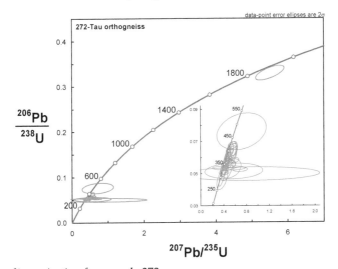

Fig. 18. Concordia projection for sample 272.

It is only obvious that at such high U-content all the grains suffered amorphization prior to the Variscan eclogite event and that during this event they underwent the *Type I* high temperature recrystallization process. The incomplete radiogenic Pb loss can be explained by fluid intervention proved by [204]Pb gain.

5. U, Th and Pb isotope systems in detrital zircons from medium grade metaquartzites and paragneisses

Most frequently the detrital zircons remain stable under crustal thermodinamic conditions because they were well selected with respect to their U-content during weathering, transport, and sedimentation. This observation is generally valid for zircons from quartzitic rocks that underwent long and possible repeated sedimentary cycles. In paragneisses, however, the material is often poorly sorted and can originate from proximal sources. The above situations will be exemplified by samples from the pre-Alpine Orliga terrane in North Dobrogea, involved in a Variscan suture as a lower plate (Balintoni et al., 2010a). The basement of the Orliga terrane has been intensely migmatized during Variscan orogenic event, fact that suggests minimum temperatures around 650-700º C. We will begin by scrutinizing the detrital zircons from the sample 336, a metaquartzite.

In this particular case, the U-content in grains is quite low (out of 72 measured grains, 52 have less than 200 ppm U). Furthermore, the U/Th ratio is typical for magmatic zircons (generally smaller than 3.5), the [206]Pb/[204]Pb ratio indicates low [204]Pb content, and the concordances are surprisingly good even for early Proterozoic or Archean ages (Fig. 19). These observations confirm the lack of the isotopic disturbances in all the grains. The age data are interpreted to represent original crystallization ages in the zircon sources with no signs of the Variscan thermotectonic events recorded by zircons. The metasediment deposition age is not older than the late Cambrian.

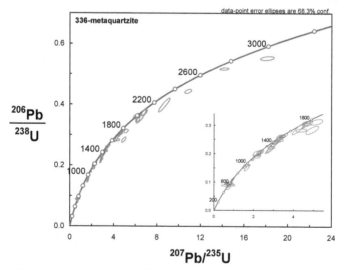

Fig. 19. Concordia projection for sample 336

Zircon Recrystallization History as a Function of the U-Content and Its Geochronologic Implications: Empirical
Facts on Zircons from Romanian Carpathians and Dobrogea

61

The zircon grains from sample 167GPS (a paragneiss) are characterized by the followings: (1) - numerous ages clustered around 300 Ma; (2) - variable U-content in zircon grains; (3) - strongly modified U/Th ratio in comparison with the sample 336; (4) - high $^{206}Pb/^{204}Pb$ in all the grains; (5) - good concordances (Fig.20).

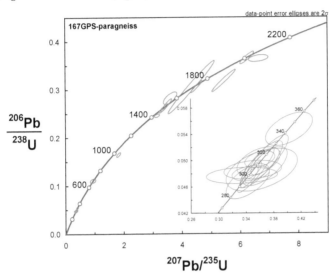

Fig. 20. Concordia projection for sample 167GPS.

The isotopic parameters for the young as well as for the old grains are presented in Table 4.

	Ages between 343.5 -294.8 Ma	Ages older than 600 Ma
U average content (ppm)	686.7	209.9
U/Th average ratio	25.8	2.4

Table 4. Comparative isotopic parameters for different sets of grains from sample 167GPS.

Considering all the data, several conclusions can be drawn. From 26 measured grains, 15 grains recrystallized during Variscan incipient melting. All these grains have an U content usually exceeding 300 ppm U. They lost all the previous radiogenic Pb, a great part of Th, and possible some U. In the same time, the recrystallized grains did not gain any ^{204}Pb and their concordances are remarkable good. These observations suggest a *Type I* recrystallization process by epitaxial migration of the interfaces between the crystalline and amorphizated parts of the grains. The grain lattices have recovered completely without fluid intervention and in presence of a melt. Apparently, ca. 300 ppm U was the boundary between the damaged and undamaged zircons.

A more complex situation is depicted by the data of sample 335.

In great lines the processes are similar to those in sample 167GPS. However, two parameters are more deteriorated in the zircons of sample 335 than in sample 167GPS: many grains gained ^{204}Pb and the ages moved away of Concordia (Fig. 21).

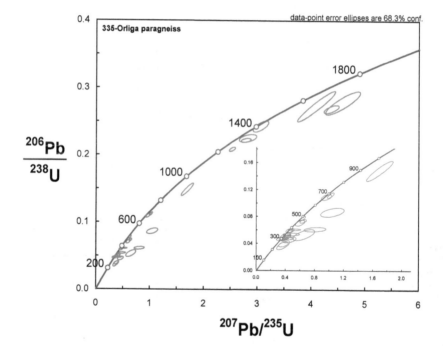

Fig. 21. Concordia projection for sample 335.

Zircon Recrystallization History as a Function of the U-Content and Its Geochronologic Implications: Empirical
Facts on Zircons from Romanian Carpathians and Dobrogea

63

These two facts suggest that fluids were extensively involved in the recrystallization processes. Several zircon grains exhibiting ages bellow 300 Ma indicate also later disturbing events.

To see again the role of the U content in zircons history we exemplify by the sample 168GPS.

With the exception of 3 grains (out of 34 analyzed grains), all the other have their U content less than 200 ppm. None of the analyzed grain was isotopically disturbed, clear evidence that at such low U content the effects of the amorphization process are indiscernible and the lattice damage do not accumulate even in geological time. All the grains show the original ages and no sign of the Variscan thermotectonic event is evident (Fig.22).

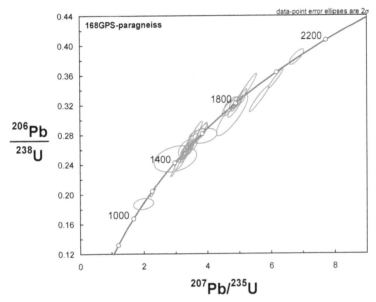

Fig. 22. Concordia projection for sample 168GPS.

6. Conclusions

According to the presented data, the boundary between accumulation of lattice damage and continuous recovery, below the critical amorphization temperature can be set at around 300 ppm U concentration. The undamaged lattices with less than 300 ppm U, sometimes show along the Concordia a spread of ages of ca. 40 Ma from which two valid weighted mean of Concordia ages can be obtained. Usually, the grains display a slight increase in the U-content and U/Th ratio toward the younger ages. The younger age is probably closer to the real crystallization age of the rock.

The migration of ages with respect to the Concordia, when the zircon grains contain over 300 ppm U, can not be explained easily in many situations. This is at least partly because some of the grains appear to lose all the radiogenic Pb at intermediate stages between well known major thermotectonic events. However, the high U-content grains can be useful in deciphering the thermotectonic history of the rocks whenever they reset dominantly around younger ages. If all the grains from a sample contain more than 300 ppm U, the initial age can be totally reset by the subsequent thermotectonic events.

Both, the magmatic and detrital zircons with over 300 ppm U record the metamorphic events, yet the low-U detrital zircon from metaquartzitic rocks do not reset throughout the crustal thermotectonic history.

Alteration in the presence of fluids strongly promotes the resetting of isotope systems as well as recrystallization even at temperatures bellow the crytical amorphization threshold. The amorphization grade is proportional with the U content, while there is little evidence for the role played by the pressure during recrystallization.

Most frequently, Th is lost together with radiogenic Pb, but U loss is by far less a common phenomenon.

7. Acknowledgements

This work was supported by a grant of the Romanian National Authority for Scientific Research, CNCS – UEFISCDI, project number PN-II-ID-PCE-2011-3-0100

8. References

Balica C., Hann H.P., Chen F., Balintoni I. & Zaharia L. 2007: The Age of the intra-Danubian Suture (Southern Carpathians, Romania). *Eos Trans. AGU* 88 (52), Abstract T31B-0476.

Balintoni I. 1997: Geotectonics of Romanian metamorphic terrains (*in Romanian*): Cluj Napoca, Ed. Carpatica, 176 p.

Balintoni I., Balica C., Ducea M.N., Chen F.K., Hann H.P. & Şabliovschi V. 2009: Late Cambrian-Early Ordovician Gondwanan terranes in the Romanian Carpathians: A zircon U-Pb provenance study. *Gondwana Research* 16, 1, 119-133.

Balintoni I., Balica C., Ducea M.N. & Stremţan C. 2011: Peri-Amazonian, Avalonian-type and Ganderian-type terranes in the South Carpathians, Romania: The Danubian domain basement. *Gondwana Research* 19, 4, 945-957.

Balintoni I., Balica C., Seghedi A. & Ducea M.N. 2010a: Avalonian and Cadomian terranes in North Dobrogea, Romania. *Precambrian Research* 182, 3, 217-229.

Balintoni I., Balica C., Ducea M.N., Zaharia L., Chen F.K., Cliveţi M., Hann H.P., Li L.Q. & Ghergari L. 2010b: Late Cambrian-Ordovician northeastern Gondwanan terranes in the basement of the Apuseni Mountains, Romania. *Journal of the Geological Society* 167, 6, 1131-1145.

Balintoni I., Balica C., Ducea M.N., Hann H.P. & Şabliovschi V. 2010c: The anatomy of a Gondwanan terrane: The Neoproterozoic–Ordovician basement of the pre-Alpine Sebeş–Lotru composite terrane (South Carpathians, Romania). *Gondwana Research* 17, 561-572.

Berza T. & Seghedi A. 1983: The crystalline basement of the Dnubian units in the Central South Carpathians: Constitution and metamorphic history. *Anuarul Instititului de Geologie şi Geofizică* LXI, 15-22.

Zircon Recrystallization History as a Function of the U-Content and Its Geochronologic Implications: Empirical
Facts on Zircons from Romanian Carpathians and Dobrogea

65

Black L.P., Williams I.S. & Compston W. 1986: Four zircon ages from one rock: the history of a 3939 Ma-old granulite from Mt. Stones, Enderby Land, Antarctica. Contrib. Mineral. Petrology 94, 427-437.

Cherniak D.J. & Watson B.E. 2000: Difusion in zircon. In: Hanchar W.M., Hoskin P.W.O. (Eds) Zircon. MSA Reviews in Mineralogy and Geochemistry 53, 113-143.

Davis D.W., Williams I.S. & Krogh T. 2003: Historical development of zircon Geochronology. In: Hanchar W.M., Hoskin P.W.O. (Eds) Zircon. MSA Reviews in Mineralogy and Geochemistry 53, 145-181.

Davis D.W. & Paces J.B. 1990: Time Resolution of Geologic Events on the Keweenaw Peninsula and Implications for Development of the Midcontinent Rift System. Earth and Planetary Science Letters 97, 1-2, 54-64.

Ewing R.C., Meldrum A., Wang L.M., Weber W.J. & Corrales L.R. 2003: Radiation effects in zircon. In: Hanchar W.M., Hoskin P.W.O. (Eds) Zircon. MSA Reviews in Mineralogy and Geochemistry 53, 387-425.

Farges F. 1994: The structure of metamict zircon: A temperature dependent EXAFS study. Physics and Chemistry of Minerals 20, 504-514.

Geisler T., Ulonska M., Schleicher H., Pidgeon R.T. & van Bronswijk W. 2001: Leaching and differential recrystallization of metamict zircon under hydrotermal conditions. Contributions to Mineralogy and Petrology 141, 53-65.

Geisler T., Pidgeon R.T., van Bronswijk W. & Kurtz R. 2002: Transport of uranium, thorium and lead in metamict zircon under low temperature hydrothermal conditions. Chemical Geology 191, 141-154.

Geisler T., Pidgeon R.T., Kurtz R. & van Bronswijk W. 2003: Experimental hydrothermal alteration of partially metamorphic terranes. American Mineralogist 88, 1496-1513.

Geisler T., Schaltteger U. & Tomascheck F. 2007: Re-equilibration of zircon in aequous fluids and melts. Elements 3 (1), 43-50.

Gruenenfelder M., Popescu G., Soroiu M., Arsenescu V. & Berza T. 1983: K-Ar and U-Pb Dating of the Metamorphic Formations and the Associated Igneous Bodies of the Central South Carpathians. Anuarul Instititului de Geologie şi Geofizică Bucureşti LXI, 37-46.

Harley S.L. & Kelly N.M. 2007: Zircon, tiny but timely. Elements 3, 13-18.

Heaman L.M., Lecheminant A.N. & Rainbird R.H. 1992: Nature and Timing of Franklin Igneous Events, Canada - Implications for a Late Proterozoic Mantle Plume and the Break-up of Laurentia. Earth and Planetary Science Letters 109, 1-2, 117-131.

Iancu V., Măruntiu M., Johan V. & Ledru P. 1998: High-grade metamorphic rocks in the pre-Alpine nappe stack of the Getic-Supragetic basement (Median Dacides, South Carpathians, Romania). Mineralogy and Petrology 63, 173-198.

Kräutner H.G., Kräutner F., Tănăsescu A. & Neacşu V. 1976: Interpretation des ages radiometique K/Ar pour les roches metamorphiques régénérées. Un exemple – les Carpathes Orientales. Anuarul Instititului de Geologie şi Geofizică Bucureşti L, 167-229.

Liégeois J.-P., Berza T., Tatu M. & Duchesne J.C. 1996: The Neoproterozoic Pan-African basement from the Alpine Lower Danubian nappe system (South Carpathians, Romania). Precambrian Research 80, 281-301.

Medaris G., Ducea M., Ghent E. & Iancu V. 2003: Conditions and timing of high-pressure Variscan metamorphism in the South Carpathians, Romania. Lithos 70, 141-161.

Meldrum A., Boatner L.A., Weber W.J. & Ewing R.C. 1998: Radiation damage in zircon and monazite. *Geochimica et Cosmochimica Acta* 62, 14, 2509-2520.

Mezger K. & Krogstad J.E. 1997: Interpretation of discordant U-Pb zircon ages: An evaluation. *Journal of Metamorphic Geology* 15, 127-140.

Murakami T., Chakoumakos B.C., Ewing R.C., Lumpkin G.R. & Weber W.J. 1991: Alpha decay event damage in zircon. *American Mineralogist* 76, 1510-1532.

Nasdala L., Wenzel M., Vavra G., Irmer G., Wenzel T. & Kober B. 2001: Metamictisaton of natural zircon: accumulation versus thermal annealing of radioactivity-induced damage. *Contributions to Mineralogy and Petrology* 141, 125-144.

Putnis A. 2002: Mineral replacement reactions: from macroscopic observations to microscopic mechanisms. *Mineralogical Magazine* 66, 5, 689-708.

Putnis C.V., Tsukamoto K. & Nishimura Y. 2005: Direct observations of pseudomorphism: compositional and textural evolution at a fluid-solid interface. *American Mineralogist* 90, 11-12, 1909-1912.

Rubatto D. & Hermann J. 2007: Zircon behaviour in deeply subducted rocks. *Elements* 3, 1, 31-36.

Salje E.K.H., Chrosch J. & Ewing R.C. 1999: Is "metamictization" of zircon a phase transition? *American Mineralogist* 84, 1107-1116.

Săbău G. & Massone H.J. 2003: Relationships among eclogite bodies and host rocks in the Lotru metamorphic suite (South Carpathians, Romania): Petrological evidence for multistage tectonic emplacement of eclogites in a medium-pressure terrain. *International Geology Review* 45, 1-38, 225-262.

Tagami T. & Shimada C. 1996: Natural long-term annealing of the zircon fission track system around a granitic pluton. *Journal of Geophysical Research-Solid Earth* 101, B4, 8245-8255.

Tilton G.R., Davies G.L., Wetherill, G.W. & Aldrich L.T. 1957: Isotopic ages of zircon from granites and pegmatites. *Transactions - American Geophysical Union* 38, 360-371.

Tromans D. 2006: Solubility of crystalline and metamict zircons: A thermodynamic analysis. *Journals of Nuclear Materials* 357, 221-233.

Wetherill, G.W. 1956: Discordant Uranium-Lead ages, I. *Transactions - American Geophysical Union* 37, 320-326.

Williams I.S. 1992: Some observations on the use of zircon U-Pb geochronology in the study of granitic rocks. *Transactions of the Royal Society of Edinburgh* 83, 447-458.

4

Recrystallization Behavior During Warm Compression of Martensite Steels

Pingguang Xu[1] and Yo Tomota[2]
[1]Japan Atomic Energy Agency, Tokai, Ibaraki,
[2]Ibaraki University, Hitachi, Ibaraki,
Japan

1. Introduction

The application of high strength-toughness-ductility structural steels is beneficial to reduce the body weight of automotives and to improve the usage efficiency of energy without any potential damage of safe and security of human beings. Grain refinement is an important fundamental research field for the development of such low alloy structural steels. The conventional thermo-mechanically controlled process (TMCP) of ferrite or ferrite-pearlite steels including severe deformation at a lower temperature of ferrite transformation and rapid cooling is usually utilized to refine the grain size down to about 5 microns. For ferrite/pearlite steels, the grain refinement through dynamic recrystallization was observed to take place at a true strain of 1.2 at 873K, and the fully recrystallized ferrite/cementite microstructure may be realized at a strain of 2.0 (Torizuka, 2005). The final grain size is dependent on the Zener-Hollomon parameter, Z, which is given by

$$Z = \dot{\varepsilon} \cdot \exp(\frac{Q}{RT}) \tag{1}$$

where $\dot{\varepsilon}$, Q, R and T refer to the strain rate, the activation energy, the gas constant, and the absolute temperature, respectively. Because the severe deformation in uni-directional rolling cannot meet the requirement on Z-value for full recrystallization, the multi-directional groove rooling process has been developed recently. Unfortunately, the groove rolling process is unsuitable to the commercial production of wide steel plates.

In various microstructure types, the martensite was expected to have a low critical strain requirement for grain refinement because the high density dislocations, the supersaturated solute carbon and the ultrafine laths are helpful to raise the stored energy. The advantage of martensite was firstly claimed (Miller, 1972) as an initial microstructure to obtain ultrafine ferrite-austenite microstructures through cold-rolling and annealing of Ni(-Mn)-C martensite steels. The cold rolling and annealing of lath-martensite was confirmed (Ameyama, 1988) much effective to make ultrafine grained microstructures of low carbon steels. Ueji et al (2002) have claimed that the formation of ultrafine grained microstructure by cold rolling followed by annealing is closely related to the fine substructures and the high density dislocations of martensite. However, the cold rolling of martensite steel requires much higher loading capacity of mills and the microcracks may occur in the surface layers and/or the side edges of martensite steel plates.

Hayashi *et al* (1999, 2002) reported that the multi-pass warm groove-rolling of low-carbon martensite could produce an ultrafine ferrite-cementite microstructure less than 1μm. Tomota *et al* proposed to realize the grain refinement by warm compression or warm rolling of martensite steel plates through the dynamic recrystallization and confirmed that the dynamic recrystallization occurs at a low critical strain during warm compression for martensite of a low-carbon SM490 steel (Bao, 2005a). This technique has been successfully employed to get ultrafine grained ferrite-cementite microstructure (Li, 2008) and ultrafine grained ferrite-austenite microstructure (Xu, 2008a). Furuhara *et al* (2007) found that an initial microstructure of high carbon martensite is preferable to reduce the critical strain for full dynamic recrystallization, showing that the cementite particles can act as hard particles to promote the dynamic recrystallization during the martensite warm deformation. Though the carbon enriched retained austenite usually transforms to martensite to improve ductility, its detailed role during warm compression was necessary to be clarified. Therefore, the neutron diffraction was applied to *in situ* investigate the precipitation of austenite and the recrystallization of ferrite in 17Ni-0.2C martensite steel during warm compression. The existence of carbon enriched austenite particles during warm compression was found to further promote the dynamic recrystallization of ferrite from lath martensite by playing a role of the hard second phase (Xu, 2008b).

In this chapter, the research progress in dynamic recrystallization and grain refinement during warm compression of martensite steels was reviewed systematically while the advantages of *in situ* neutron diffraction were emphasized as a powerful beam technique suitable for clarifying the microstructure evolution during various thermo-mechanically controlled processes.

2. Dynamic recrystallization during warm compression in low alloy martensite steel

The dynamic recrystallization is one of the effective ways to refine microstructure to improve strength-toughness balance. Torizuka *et al* (2005) carried out a systematic study on warm compression and warm rolling for a conventional low-carbon ferrite/pearlite steel, realized a submicron ferrite-cementite microstructure through continuous dynamic recrystallization during heavy deformation up to a true strain of 3~4 and successfully produced the ultrafined grained high strength wires, rods and even steel plates with 300mm in width. However, it is difficult to apply such severe deformation with a high Z-value to commercial production of wider plates or sheets.

The average grain size of dynamically recrystallized microstructure was found to be dependent on Z-value and independent on the initial micrsotructure type such as ferrite or ferrite-pearlite or martensite (Ohmori, 2002; Ohmori, 2004; Bao, 2005a; Bao, 2005b). Though the initial grain size and the compressive strain in cases of ferrite or ferrite-pearlite initial microstructure have no direct effect on the average grain size of final ferrite microstructure, it is not clear about the effects of initial austenite grain size, strain and pre-tempering treatment on the dynamic recrystallization and the final grain size during the warm compression of martensite. Li *et al* (2008) investigated the effects of initial microstructure, deformation strain and pre-tempering on dynamic recrystallization of ferrite from lath martensite by using a commercial low-carbon steel, JIS/SM490 (mass%: 0.16C-1.43Mn-

0.41Si-0.014P-0.004S-0.01Cu-0.027Al-0.028N). The transformation temperatures measured by dilatometry at a heating and cooling speed of 5K/s were 1010K for Ac1, 1133K for Ac3 and 895K for Ar1, respectively (Bao, 2005a). Cylindrical compressive specimens with ϕ4mm×6mm were spark cut from the 15×15mm martensite steel bars obtained by water quenching after solid solution treatment at 1273K for 3.6ks. The specimens were heated up to 873K or 923K, held there for 1s, compressed by true strain ε = 0.3, 0.55 or 0.7 at $1.7×10^{-3}$ s^{-1}, and then quenched into water. Some specimens were pre-tempered at 873K or 923K for 3.6ks before the warm compression. The deformed specimens were cut along the longitudinal direction and the central area of the sectioned plane was observed with a field emission scanning electronic microscope (FE-SEM).

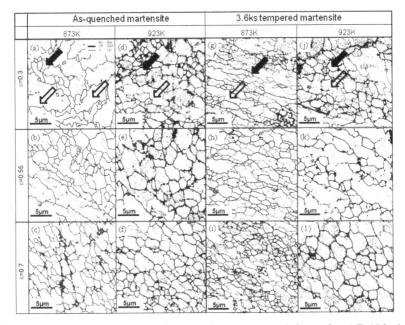

Fig. 1. Boundary misorientation mappings for the specimens deformed at $1.7×10^{-3}$ s^{-1} where the initial martensite was quenched from 1273K.

Fig.1 summarized the boundary misorientation mappings for the specimens warm compressed at $1.7×10^{-3}$ s^{-1}, where the solid arrows marked the recrystallized ferrite grains surrounded only by high angle grain boundaries (here called as Type I grains, according to Torizuka's definition (2005)) and the open arrows marked the recrystallized ferrite surrounded partially by high angle boundaries (here called as Type II grains). The Type I grains can be observed after the true strain 0.3 in all cases, revealing much lower critical strain for the initiation of dynamic recrystallization of low carbon martensite microstructure. This is surprizingly low because the critical strain is higher than 1.0 for the conventional ferrite-pearlite initial microstructure (Torizuka, 2005). The Type II grains equaxilize gradually and its size becomes almost equal to that of Type I grains with increasing of strain. It is also found that the grain size of Type I grains at a lower temperature (873K) deformation is smaller than that at a higher temperature (923K), revealing that the grain size is dependent on the Z-value. Moreover, compared with (c) and (i), it can be found that the 3.6ks prior tempering at 873K leads to the increase of Type I grains and smaller Type II grains, suggesting the prior tempering promote the dyanmic recrystallization. However, compared with (f) and (l), the 3.6ks prior tempering at 923K does not acclerate the grain refinement process. The observation of initial micorstructure before warm compression shows that the pretempering at 873K enables the dispersive precipitation of cementite particles and the dislocationes in martensite do not disappear so much, while the pretempering at 923K reduces the dislocation density and increases the size of cementite particles (Li, 2008). Consequently, the acceleration of dynamic recrystallization after the lower temperature pre-tempering (873K) is mostly related to the dispersive precipitation of cementite particles, while the delay of dynamic recrystallization after the higher tempatrue pretempering (923K) is mostly related to the decrease in dislocation density.

3. Dynamic recrystallization and austenite precipitation during warm compression in high nickel martensite steels

While the submicron ultrafine grained ferrite/cementite microstructure shows the limited ductility, the submicron ultrafine grained austenite/ferrite duplex microstructure possesses up to 1000MPa tensile strength and 25% uniform elongation (Tomota, 2008). Because the austenite can evidently improve the tensile ductility through the transformation induced plasticity (TRIP) effect, it leads to higher possibility for industrial applications. As mentioned in Section 1, warm compression or rolling of martensite microstructure with initial fine substructures and the high density dislocations of martensite can be utilized to accelerate the recrystallization and refine grains at a low critical strain (Bao, 2005a and 2005b). It was suspected that the competition of austenite precipitation and ferrite recrystallization played an important role to obtain ultrafine grained structures (Enomoto, 1977). However, the effects of austenite amount and its thermomechanical stability related to carbon concentration on the dynamic recrystallization of ferrite and the formation of ultrafine grain microsructure have not been clarified (Maki, 2001).

Xu et al investigated the effects of austenite precipitation and the carbon enrichment in austenite on the dynamic recrystallization during warm compression by using 18Ni and 17Ni-0.2C (mass%) martensite steels (Xu, 2008a) where 843K and 773K were chosen as the pre-tempering temperature according to the phase diagrams of Fe-Ni and Fe-Ni-C alloys, respectively. Cylindrical samples with 6.5 mm in length and 4mm in diameter for compression tests were prepared by spark cutting and surface grinding. The samples were heated up to the deformation temperature at a heating rate of 5K/s and then deformed at 8.3×10^{-4}/s followed by water quenching. The longitudinally sectioned specimens were electrochemically polished to avoid the stress-induced martensite transformation of austenite.

3.1 SEM microstructure observation and EBSD microstructure characterization

Fig.2 showed the different pre-tempered 17Ni-0.2C microstructures before and after warm compression at 773K. Different from the non-tempered martensite (Fig. 2(a)), small particles can be found in the tempered martensite matrix (marked by circle in Fig. 2(b)), and these particles are coarser (marked by circle in Fig.2(c)) after long-time pre-tempering. In the specimen warm compressed without any pre-tempering, ultrafine equiaxed grains (marked by white arrow) predominate the SEM microstructure while some elongated grains (marked by gray arrow) related to recovery can be also found. In contrast, the warm deformed specimen with 3.6ks pre-tempering shows a fully equiaxed microstructure, and the ferrite grains are much finer. However, the warm compression after 36ks pre-tempering results in a partially equiaxed microstructure, revealing the long-time pre-tempering retards the ferrite recrystallization evidently and a larger compressive strain is necessary to achieve full recrystallization.

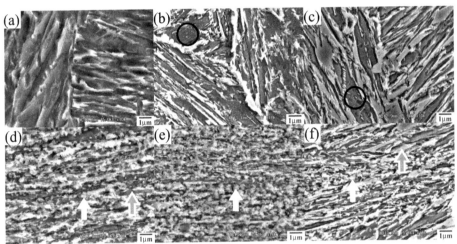

Fig. 2. SEM microstructures of 17Ni-0.2C steels before (a, b and c) and after (d, e and f) warm compression at 773K, ε=0.6, 8.3×10⁻⁴/s. (a, d) non-tempered; (b, e) 3.6ks tempered; (c, f) 36.0ks tempered. The compression axis is along the vertical direction. (Xu, 2008a)

Fig.3 showed the electron back-scattering diffraction (EBSD) maps of 18Ni steel microstructures before and after warm deformation, where the compression axis is along the vertical direction. The volume fraction of the retained austenite amount in the non-tempered microstructure is less than 1%, and increases hardly after the 3.6ks pre-tempering.

In the ε=0.6 warm compressed specimen without pre-tempering, the equiaxed ferrite grains are dominative although many grains are still with small misorientation; in case of 3.6ks pre-tempering, the equiaxed ferrite grains are much few and the average grain size is a little coarser, suggesting that the decrease in dislocation density in the initial microstructure by pre-tempering brings clearly negative influence on the dynamic recrystallization.

After a larger strain warm compression (ε=0.8), the fully equiaxed ferrite grains can be obtained in both cases and their average grain sizes are approximately equal. Although blocky austenite grains are not observed in the microstructures after warm compression, the local regions with a confidence index of less than 0.2 (marked by circles) are related to the metastable austenite at high temperature, partially transformed to martensite /austenite islands during rapid cooling to room temperature for the sample preparation. Considering that such conventional microstrure quenching(freezing) process is not so ideable, the *in situ* microstructure evaluation involving neutron diffraction (see Section 4) is much important for investigation on the high temperature microstructure evolution.

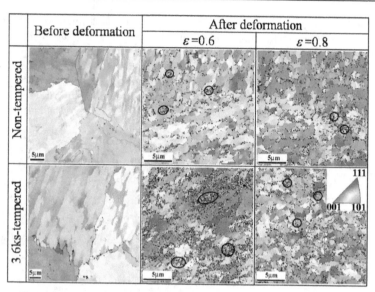

Fig. 3. Crystallographic orientation characteristics of 18Ni martensite steel before and after warm compression at 843K, 8.3×10^{-4}/s. (Xu, 2008a).

Fig.4 gives the EBSD maps of 17Ni-0.2C microstructures before and after warm compression. The austenite volume fraction is plotted in Fig.5. For the non-tempered specimen, the austenite amount increases much after deformation (from 9.5% to 17%) and its average grain size of austenite decreases evidently because of the precipitation of ultrafine grained austenite during warm deformation. After warm deformation, the equiaxed ferrite grains in the local regions with bulky austenite (marked by circle) are with large misorientation but those in the regions with no bulky austenite (marked by reactangle box) are with small misorientation, showing that the ferrite dynamic recrystallization is related to the existence of bulky austenite grains.

The 3.6ks pre-tempering leads to an clear increase of austenite amount. After the ε=0.6 warm compression, the ferrite-austenite duplex microstructure becomes fully equiaxed and the volume fraction of austenite increases from 18% before deformation to 36.8% after deformation and the average size of austenite grains decreases from 1.53μm before deformation to 0.94μm after deformation, which also proves that the warm deformation promotes the precipitation of ultrafine grained austenite. Meanwhile, the ferrite grains are refined to 0.85μm after deformation, smaller than that of 3.6ks pre-tempered 18Ni steel after warm compression at a strain of 0.8. It reveals that the higher Z-value (17Ni-0.2C, 773K, $Z=1.21 \times 10^{14}$ s^{-1}; 18Ni, 843K, $Z=4.55 \times 10^{12}$ s^{-1}) and the high carbon content are helpful to refine the fully recrystallized microstructure.

When the pre-tempering time prolongs to 36.0ks, the austenite amount approaches 33.8% after the isothermal tempering treatment. After the ε=0.6 warm deformation, the ferrite grains are elongated and coarse, showing that the dynamic recrystallization is delayed evidently. Meanwhile, the austenite amount measured at room temperature decreases abruptly to 13.8%. According to the phase diagram, the equilibrium microstructure at

773K consists of ferrite, austenite and cementite. The austenite amount after long-time pre-tempering in Fig.5 increases evidently due to the austenite precipitation at an elevated temperature below the martensite reverse transformation starting temperature, which is consistent with the previous reports for a Fe-Ni martensite alloy (Enomoto, 1977) and for a 18Ni-12Co-4Mo maraging steel (Moriyama, 2001). The decreasing austenite amount after warm compression is related to the weakening of austenite thermal stability caused by the carbon depletion in austenite after cementite precipitation. Since the cementite particles are mostly in nanometer scales, TEM observations are necessary.

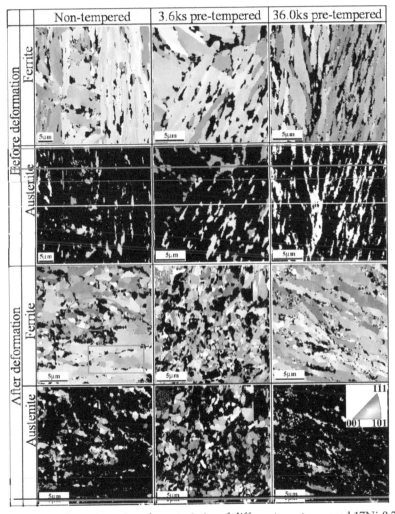

Fig. 4. Crystallographic orientation characteristics of different pre-tempered 17Ni-0.2C martensite steel specimens before and after warm compression at 773K, 8.3×10⁻⁴/s. The compression axis is along the vertical direction (Xu, 2008a).

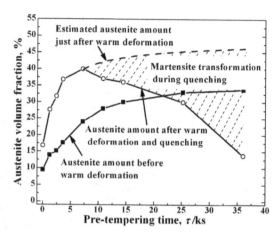

Fig. 5. Change in the austenite amount of 17Ni-0.2C steel by pre-tempering and warm deformation at 773K according to EBSD phase mapping statistics (Xu, 2008a).

3.2 TEM microstructure observation and EDX composition analysis

Fig.6 shows the bright field images for the non-tempered (a) and pre-tempered (b, c) microstructures of 17Ni-0.2C steel after warm compression. In the case of no pre-tempered condition, large ferrite grains near the austenite particles can be found with high density dislocations and some new grains with no intragranular dislocations form in these regions. It is found in Fig. 6(a) that the high angle grain boundaries surrounded an individual ferrite grain. In the warm compressed specimen after 3.6ks pre-tempering, the ultrafine equiaxed ferrite and austenite grains can be found frequently (Fig. 6(b)), suggesting that the warm deformation promotes the austenite precipitation and the ferrite dynamic recrystallization. The misorientation analysis confirms that the grain boundaries between these grains and their neighbors are of high angle boundaries, exhibiting that the dynamic recrystallization of ferrite can be confirmed really to take place during the warm deformation.

In case of the 36ks tempering, the elongated martensite laths within low dislocation density are clearly observed, which is related to the recovery of martensite. It is also found that the cementite precipitates preferably around the carbon-enriched austenite grains located at the martensite lath boundaries rather than within the martensite laths. The warm deformation accelerates the carbon diffusion and promotes the dynamic precipitation of spherical cementite particles within the martensite laths (Fig.6(c)).

For a pre-tempering time longer than 7.2ks, the carbon partitioning from martensite to austenite is gradually replaced by the cementite precipitation. Because of the precipitation of cementite particles and the increase of austenite amount, the average carbon concentration in austenite must decrease, finally lower than that in the 3.6ks tempered and deformed specimen. As a result, the thermal stability of austenite decreases, some austenite grains transform to martensite upon cooling after deformation and the fresh nanometer-scale martensite particles form around the cementite and austenite particle (Fig.6(c)). Hence the austenite amount after rapid cooling to room temperature becomes smaller (see Fig.4). The low density dislocations and no carbon-enriched austenite lead to the delayed dynamic recrystallization in the 36ks pre-tempered and warm compressed specimen when compared with the 3.6ks pre-tempered and warm compressed specimen.

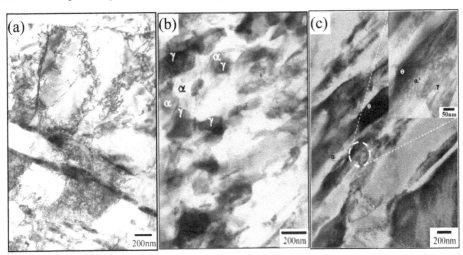

Fig. 6. Transmission electron microscopic microstructures of different pre-tempered 17Ni-0.2C steel after warm compression at 773K, 8.3x10⁻⁴ /s: (a) non-tempered; (b) 3.6ks pre-tempered; (c) 36ks pre-tempered.

According to the Fe-Ni and Fe-Ni-0.2C equilibrium phase diagrams, the nickel-rich austenite should appear at 843K in 18Ni and the nickel-rich and carbon-rich austenite and cementite should appear at 773K in 17Ni-0.2C. The EDX composition analysis proves that the metastable austenite (or martensite) is nickel-enriched, for example, the nickel content is about 20-26mass% (austenite) versus 15-16mass% (ferrite) in 17Ni-0.2C. Though the EDX analysis can not provide a reliable carbon concentration in constituent phases, the austenite in the 3.6ks pre-tempered 17Ni-0.2C after warm compression should be carbon-enriched because (a) no cementite particles are found in such microstructure, (b) the stability of austenite phase is much higher than that in 18Ni, (c) the diffusion of carbon is more rapid than that of nickel and (d) the interstitial solubility of carbon is much higher in austenite than in ferrite. Considering that the 3.6ks pre-tempering decreases the dislocation density of martensite substructures, the accelerate dyamic recrystallization is mainly related to the formation of a large amount of carbon-enriched austenite.

3.3 Effect of carbon-enriched austenite on dynamic ferrite recrystallization

Generally, the retained austenite at room temperture is softer than the corresponding martensite. If there is no *in situ* quantitative micromechanical data of two duplex microstructure at elevated temperature, it will be much difficult to clarify why the existence of carbon-enriched austenite promotes the dynamic ferrite recrystallization during warm compression. Fortunately, the *in situ* neutron diffraction study about 3.6ks pre-tempered 17Ni-0.2C steel (Xu, 2008b) showed that the heterogeneous deformation occurs during warm compression at 773K and the ferrite matrix is subject to larger plastic deformation than the carbon-enriched austenite, i.e. the former is softer than the latter. Moreover, it was found that before and after approaching the critital strain ε=0.13, the ferrite (hkl) reflections showed clear difference among their relative integrated intensities because of the occurrence of dynamic ferrite recrystallization. In addition, the splitting of austenite reflection peaks and its disappearance during warm compression obtained by a neutron diffraction study (Xu, 2008b) help us to understand the change in carbon concentration of austenite during the warm compression. Consequently, the influencing mechanism of carbon-enriched austenite on the heterogeneous deformation and the ferrite-austenite duplex microstructure evolution can be summarized as shown in Fig.7.

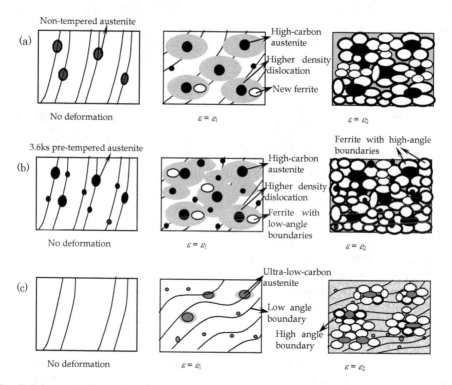

Fig. 7. Schematic illustration for microstructure evolution of martensite steels with different carbon contents during warm deformation, where the compression axis is along the vertical direction and $\varepsilon_1 < \varepsilon_2$: (a) 17Ni-0.2C, non-tempered, at 773K; (b) 17Ni-0.2C, 3.6ks pre-tempered, at 773K; (c) 18Ni, at 843K. (Xu, 2008a)

For the carbon-added high-nickel steel, the amount of carbon-enriched austenite increases due to the proper pre-tempering and the austenite particles becomes harder than the pre-tempered martensite (or recovered ferrite matrix) during warm compression. Hence the larger local plastic flow takes place in the regions around austenite particles. When the local plastic strain increases to a critical value ($\varepsilon = \varepsilon_1$), the recrystallized ferrite grains preferably nucleate in these local regions near the austenite particles according to the related recrystallization model (Doherty, 1997). When the true strain increases to another critical value ($\varepsilon = \varepsilon_2$), all the ferrite grains are equiaxed and fully recrystallized (Fig. 7(b)).

In case of no pre-tempering, the austenite amount is relatively less than that in case of 3.6ks pre-tempering and the carbon content in austenite is lower than that in the latter. Although the warm compression promotes the austenite precipitation and the carbon enrichment in austenite, the plastic deformation partitioning between austenite particles and recovered ferrite matrix is a little weaker. Consequently, at the true strains $\varepsilon = \varepsilon_2$, the recrystallized ferrite amount is correspondingly less than that in the case of 3.6ks pre-tempering (Fig. 7(a)).

When the pre-tempering time is extended to more than 14.4ks, the precipitation of cementite particles leads to relative carbon depletion in austenite, and the plastic deformation partitioning between the austenite and the recovered ferrite matrix is no longer apparent. Hence the dynamic recrystallization of ferrite is retarded markedly (Maki, 2001). The coarse austenite grains formed during long-time pre-tempering are metastable because of low carbon concentration, easy to deform during warm compression, and then transform to martensite (or ferrite matrix) upon rapid cooling to room temperature.

For the ultra-low-carbon high-nickel steel, the plastic deformation partitioning is relatively weak since the carbon-enriched austenite is impossible to form and martensite is easy to recover. Moreover, the dislocation density in the martensite matrix is lower than that in carbon-added high-nickel steel and the block is also coarser. As a result, a larger compressive strain is needed for the full recrystallization (Fig. 7(c)).

As a summary, the following conclusions should be mentioned. The carbon addition is beneficial to reduce the critical strain for full recrystallization of high-nickel martensite steels during warm compression. The increment of carbon-enriched austenite volume fraction accelerates the dynamic recrystallization of ferrite through plastic deformation partitioning in the 17Ni-0.2C steel. Proper pre-tempering promotes the precipitation of the carbon- and nickel-enriched austenite, and then promotes the dynamic recrystallization in the 17Ni-0.2C steel. On the other hand, long-time tempering leads to the carbon depletion in austenite so as to delay the dynamic recrystallization. In the 18Ni steel, a larger warm compression strain is required for full recrystallization during warm compression, mainly because of no carbon-enriched austenite.

4. Dynamic microstructure evolution in high nickel martensitic steel during warm compression studied by *in situ* neutron diffraction

As mentioned in above sections, a large strain and a high Zener-Hollomon value may be unnecessary to obtain ultrafine grained microstructures if the martensite initial microstructure is employed during annealing after cold working (Miller, 1972; Ameyama, 1988) or during warm deformation (Hayashi, 1999, Bao, 2005a). The microstructural observation of warm compressed specimens with different amounts of pre-existing austenite has shown that a dynamically recrystallized microstructure can be obtained in a 17Ni-0.2C martensite steel at a small strain ($\varepsilon \leq 0.6$). In order to make clear the microstructure evoluton and to investigate the effect of the austenite particles on dynamic recrystallization, the *in situ* Time-Of-Flight (TOF) neutron diffraction for the as-quenched 17Ni-0.2C steel during warm compression was carried out using the ENGIN-X neutron diffractometer at ISIS, Rutherford Appleton Laboratory.

Since the warm deformation at a temperature just below the austenite transformation starting temperature accelerated the austenite transformation significantly, a little low temperature (773K) was selected in order to investigate the dynamic recrystallization of ferrite. Here, 3.6ks pre-tempering treatment at 773K was carried out to obtain about 18vol% austenite in the initial microstructure. Cylinder specimens with 8mm diameter and 20mm length were prepared by spark cutting and surface grinding.

A 100kN hydraulic loading rig attached with a radiant furnace with a control error of ±1K was employed to realize the thermomechanically controlled process (TMCP). The specimen was heated up to 773K and neutron diffraction spectrum during the isothermal holding was collected repeatedly with 1min acquisition time to investigate the effect of the tempering process on the microstructure evolution. After 600s isothermal holding at 773K, the specimen was compressed at a strain rate of 8.3×10^{-4} s^{-1} and the neutron diffraction spectra were acquired at 773K. In order to decrease the experimental error, each neutron spectrum was summed with the consecutive one for the purpose of applying the Rietveld refinement.

Because it was difficult to distinguish the diffraction peaks of ferrite and martensite using the relative weak neutron spectra, the tempered/deformed martensite and the recrystallized ferrite during warm compression are here simply designated as the ferrite matrix. The austenite volume fraction was determined by the Rietveld refinement using the General Structure Analysis System (GSAS) software package (Larson, 2004), taking all diffraction peaks measured into consideration. The texture indexes (Bunge, 1982) of warm deformed austenite and ferrite were evaluated by the spherical harmonic preferential orientation fitting with an assumption of cylindrical sample symmetry, and the series was truncated at a maximum expansion order of $l_{max}=8$. Single peak fitting with the third TOF profile function (Larson, 2004) was employed to obtain the integrated intensities and the lattice spacings of (hkl) peaks. The preferred orientations and the lattice strains of ferrite and austenite were also analyzed by these integrated intensities and lattice spacings, respectively.

4.1 Microstructure evolution during isothermal holding

Fig.8 gives the axial neutron spectrum acquired in 600s at room temperature and the corresponding Rietveld refinement result where the small residual error reveals that the profile fitting quality is good enough. The austenite amount of the 3.6ks pre-tempered 17Ni-0.2C martensite steel is about 21.0±0.3%, a little higher than that measured by EBSD technique for an electrochemically polished sample (about 18%), which is possibly related to the weak stability of austenite, *i.e.* martensite transformation near surface (Chen, 2006). Actually, the neutron diffraction, as an important materials characterization technique different from the X-ray and electronic diffraction, enables to measure a large sample with a large gauge volume and to get bulk average information with high statistics and to evaluate the microstructure evolution under various environments, especially for multiphase materials.

Fig.9 shows the change in austenite mass fraction during isothermal holding at 773K. The austenite amount increases slowly and results in about 1.5% increment after 600sec isothermal holding. This precipitation speed is consistent with that estimated by the microstructure observation where the austenite amount increases from 9.5% before tempering to about 18% after 60min tempering at 773K (Xu, 2008a). These results also reveal that the neutron diffraction technique is suitable to evaluate the austenite precipitation during isothermal holding.

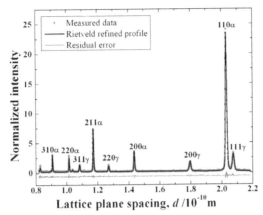

Fig. 8. Neutron diffraction spectrum of 3.6ks pre-tempered 17Ni-0.C martensite steel obtained at room temperature and its Rietveld refinement result taking all diffraction peaks measured into consideration. (Xu, 2008b)

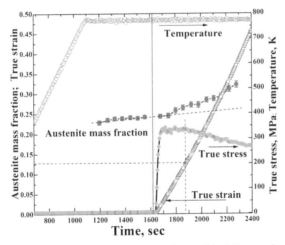

Fig. 9. Change in austenite fraction during 773K isothermal holding and warm compression.

4.2 Change in austenite diffraction spectra during warm compression

Fig.9 gives the changes in austenite fraction and true strain monitored with a high-temperature extensometer during warm compression as a function of time, showing that the austenite amount increases gradually with increasing compressive strain, i.e. from 24% at $\varepsilon=0.0$ to 32% at $\varepsilon=0.33$. Compared with the isothermal holding test, it is clear that the isothermal compression accelerates the austenite precipitation. In addition, the flow stress approaches the maximum value at a true strain about $\varepsilon=0.13$, and then gradually decreases.

Accordingly, the stress-strain curve is divided into two parts: (1) Region I, work hardening; (2) Region II, work softening.

Fig.10 compares the (111) diffraction peaks of austenite obtained in the axial direction at different loading steps during the warm compression at 873K. Though the austenite (111) peak intensity at ε=0.05 has almost equal to that at ε=0.0, the (111) lattice plane spacing at ε=0.05 (d_{111} =0.208428±0.000013 nm) is smaller than that at ε=0.0 (d_{111}^0 =0.208826±0.000007 nm). In other words, the lattice strain of austenite, $\varepsilon_{111} = (d_{111} - d_{111}^0) / d_{111}^0$, is about -1906×10⁻⁶, which is comparable with the elastic strain of the specimen estimated by $\varepsilon = \sigma / E$ =-1781×10⁻⁶ where the external stress σ is about 330MPa. Here the high-temperature Young's modulus E is taken as 185GPa (Sawada, 2005). That is to say, the (hkl) peak shifts of austenite related to the lattice strain are mostly due to the external stress.

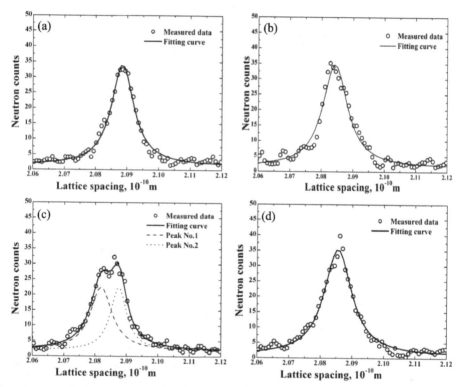

Fig. 10. Austenite (111) peaks and fitting results: (a) non-splitting, ε=0.0; (b) peak shift, ε=0.05; (c) peak splitting at ε=0.15; (d) disappearance of peak splitting at ε=0.33.

As the true strain increases to 0.15, the external stress increases only a little bit (see Fig.9), so that the amount of further peak shift is very limited. After this stage, the decreasing external stress with increasing true strain leads to a smaller peak shift in the opposite sense. On the other hand, the austenite (111) peak at ε=0.15 splits into two peaks with lower peak intensities (see Fig.10c). However, this peak splitting gradually disappears by ε=0.33, by which stage there is again a single peak with a higher intensity. Considering the peak intensities of such neutron spectra were not strong, the third TOF profile function (Larson, 2004) was employed to fit the austenite (111) peaks. The splitting of austenite (111) peak suggests that newly precipitated austenite possesses lower carbon and nickel contents from those of the pre-existing austenite. Then, the chemical composition of austenite gradually becomes uniform with increasing compressive strain.

4.3 Change in ferrite diffraction spectra during warm compression

Fig.11 compares the (110) diffraction peaks of ferrite obtained in the axial direction at different loading steps at 773K. The (110) lattice plane spacing at ε=0.05 (d_{110} =0.203425±0.000007 nm) is smaller than that at ε=0.0 (d^0_{111} =0.203780±0.000007 nm). In other words, the lattice strain of ferrite, $\varepsilon_{110} = (d_{110} - d^0_{110})/d^0_{110}$, is about -1742×10⁻⁶, lower than that of austenite ε_{111}. The hkl-specific elastic moduli E^K_{hkl}, calculated by the Kröner model, are 247.9 GPa for the austenite along the [111] direction and 225.5 GPa for the ferrite along the [110] direction (Hutchings, 2005), respectively. Therefore, the austenite is subject to a higher average phase stress than the ferrite matrix at the beginning of warm compression, i.e. the pre-existing austenite is harder than the ferrite matrix. In comparison with diffraction peaks obtained at ε=0.0, all (hkl) ferrite peaks show clear peak shifts at ε=0.05 due to the external stress about 330MPa. The ferrite (110) peak intensity decreases substantially and the ferrite (211) peak intensity increases a little. When the strain increases to ε=0.15, the ferrite (110) peak intensity decreases slowly but the ferrite (211) and (200) peak intensities decrease significantly.

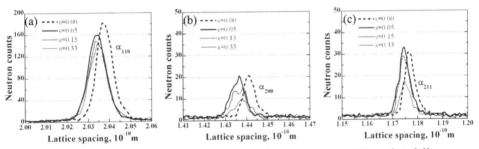

Fig. 11. Comparison of ferrite axial diffraction peaks of the 17Ni-0.2C steel at different compressive strains: (a) ferrite (110) peak; (b) ferrite (200) peak; (c) ferrite (211) peak. (Xu, 2008b)

Fig.12 shows the change in integrated intensities of different diffraction peaks with increasing of the compressive strain, obtained by single peak fitting of neutron spectra measured both in the axial and the radial directions of cylinder specimen. Before the compressive strain increases to 0.13, i.e. before the true stress reaches the maximum value, the ferrite (110) integrated intensity obtained in the axial direction decreases rapidly but that obtained in the radial direction decreases slowly. When the compressive strain is beyond ε=0.13, the decrease in ferrite (110) intensity in the axial direction becomes much slower while the decrease in the radial direction accelerates markedly. For the ferrite (200) and (211) peaks, the integrated intensities obtained in the axial direction increase slowly before ε=0.13 and then decrease evidently beyond ε=0.13, but such data obtained in the radial direction show little change during the warm compression. Based on these changes in ferrite crystallographic orientation, the strain corresponding to the maximum true stress can be regarded as the onset strain for dynamic recrystallization of ferrite from martensite.

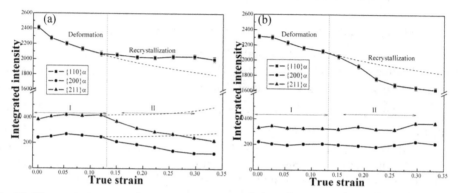

Fig. 12. Change in integrated intensity of ferrite in 17Ni-0.2C steel during warm compression: (a) obtained from the axial neutron spectra; (b) obtained from the radial neutron spectra. Regions I and II correspond to the ferrite deformation and the ferrite dynamic recrystallization, respectively. (Xu, 2008b)

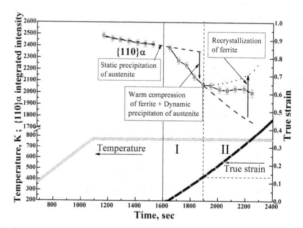

Fig. 13. Illustration for the microstructure evolution during isothermal holding and warm compression, based on the change in integrated intensity of ferrite (110) peak obtained in the axial direction.

Fig.13 illustrated the microstructure evolution based on the change in integrated intensity of ferrite (110) peak acquired in the axial direction during the isothermal holding and warm compession of 17Ni-0.2C martensite steel with 3.6ks prior tempering treatment. During the isothermal holding, the ferrite (110) integrated intensity decreases slowly due to the static precipitation of austenite. During the warm compression in the Region I, the warm compression of ferrite and the dyanmic precipitation of austenite accelerate the reduction of the ferrite (110) integrated intesity and form a clear deviation from the change trend of ferrite (110) integrated intensity in during the isothermal holding; when the strain surpasses the critical strain 0.13, the compessed ferrite grains begins to recrystalize, which leads to an evident increase of ferrite (110) integraed intensity as marked by the dotted line if there is no other microstructure evolution. The co-existence of the ferrite recrystallization as marked by the dotted line and the ferrite deformation as marked by the dashed line (which is partially related to austenite precipitation) means that the so-called ferrite dyamic recrystallization really takes place. In other words, the microstructure evolution in the Region II is involved in the dyanamic ferrite recrystallization and the dynamic austenite precipitation, and their compitition effect is believed to result into the slow decrease in the ferrite (110) integrated intensity.

4.4 Effect of the existence of austenite grains on dynamic recrystallization of ferrite

Following the dynamic recrystallization, the flow stress decreases gradually to a stable stress. Since the external loading and the composition change in constituent phases both affect the lattice plane spacing, it is difficult to compare the lattice compressive strains directly during warm compression beyond ε=0.05. Considering that the plastic deformation changes the texture as described above, the texture indexes of the two constituent phases can be used to make an indirect comparison between the plastic strains of austenite and ferrite.

According to Bunge's definition (Bunge, 1982), the texture index J is a parameter to characterize the sharpness of the texture by the integral of the square of the texture function $f(g)$, without considering the details of the crystallographic orientation distribution:

$$J = \oint [f(g)]^2 dg = \sum_{l,\mu,\nu} \frac{1}{2l+1} \left| C_l^{\mu\nu} \right|^2 \qquad (2)$$

where l, μ and ν are the series expansion orders and $C_l^{\mu\nu}$ is the corresponding expansion coefficient. For a random texture, J=1.0 and for an ideal texture of single orientation, $J \to \infty$. Generally, heavier plastic deformation leads to sharper texture. In addition, the TOF neutron spectrum measurement covers a wide range of lattice plane spacings which is partially equivalent to measuring one reflection over a wide range of sample orientations. Therefore, the texture indexes calculated from the TOF neutron spectra by the GSAS software package (Larson, 2004) can be employed to compare the plastic strains of austenite and ferrite.

As shown in Fig.14, the texture indexes of the ferrite phase are evidently larger than those of austenite in both the axial and radial directions while the difference between the two directions is related to the details of the crystallographic orientation distribution. This suggests that the plastic deformation occurs preferentially in the ferrite matrix during warm compression.

The different peak shifts in Section 4.2 & Section 4.3 reveal that the austenite grains are harder than the ferrite grains. The texture index of the ferrite matrix is larger than that of austenite, suggesting that the plastic deformation occurs preferentially in the ferrite matrix. According to the related recrystallization literature (Doherty, 1997), the recrystallized ferrite grains nucleate preferentially in the heterogeneously deformed regions near large hard particles. Therefore, the existence of carbon-enriched austenite is able to accelerate the dynamic recrystallization in the ferrite matrix.

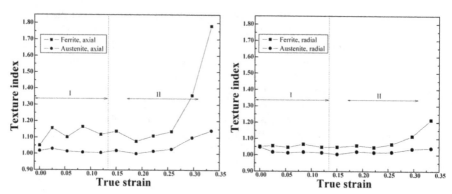

Fig. 14. Change in texture indexes of austenite and ferrite during warm compression: (a) obtained from the axial direction; (b) obtained from the radial direction. (Xu, 2008b)

The TOF neutron diffraction experiment helped us to understand the austenite precipitation and dynamic recrystallization behaviors of the 17Ni-0.2C martensite steel during isothermal warm compression after isothermal holding at 773K, and the main results were as follows: (1) the warm deformation at 773K accelerated the austenite precipitation in 17Ni-0.2C martensite steel. Splitting of the austenite (111) peak was found to occur and then disappear during warm compression. The splitting is ascribed to the different carbon concentrations in the dynamically precipitated austenite and the pre-existing austenite, and the disappearance of splitting is related to carbon homogenization at larger strain. (2) The austenite in the 17Ni-0.2C martensite steel is harder than the ferrite matrix at 773K. Heterogeneous deformation occurs preferentially in the ferrite matrix, leading to the acceleration of dynamic recrystallization.

5. Summary and future works

The low-carbon martensite steel and the high-nickel martensite steels have been warm compressed with and without prior tempering treatment to realize the microstructure refinement based on dynamic recrystallization at lower Zener-Hollomon parameter Z. Because of the lower critical strain for fully recrystallization compared with the warm deformation of ferrite or pearlite/ferrite, the warm deformatioin of martensite may be employed to the production of future ultrafine grained multiphase steels. The carbon addition promotes the formation of hard second phase particles such as cementite and austenite, which accelerates the ferrite recrystallization through the formation of local high strain regions near the hard particles during warm deformation.

Neutron diffraction has been applied as a powerful tool to investigate the microstructure evolutions of bulk materials during tensile/compressive deformation, heating/cooling and under other specific environmental conditions (te Velthuis, 1998; Tomota, 2005; Xu, 2006a; Xu, 2006b). Recently, TOF (*hkl*) multiple reflection spectra obtained by neutron diffraction have been analyzed to evaluate the crystallographic textures during forward and reverse diffusional phase transformations (Wenk, 2007) and the preferred orientations of ferrite and austenite in a 0.2C-2Mn steel before and after hot compression (Xu, 2009). The *in situ* microstructure and texture evolution during thermomechical controlled process will be studied further in order to well optimize the multiphase microstructure (Xu, 2012).

6. Acknowledgments

The authors thank Dr. Y. Adachi at National Institute for Materials Science, Japan for his support on the TEM microstructure observation. They also appreciate Dr. E.C. Oliver at ISIS Facility, Rutherford Appleton Laboratory, United Kingdom for his support on the neutron diffraction.

7. References

Ameyama, K.; Matsumura, N. & Tokizane, M. (1988). Ultrafine Austenite Grains Obtained by Thermomechanical Processing in Low and Medium Carbon Steels. *Journal of the Japan Society for Heat Treatment*, Vol.28, no.4, pp.233-240, ISSN 0288-0490.

Bao, Y.Z.; Adachi, Y.; Toomine, Y.; Suzuki, T.; Xu, P.G. & Tomota, Y. (2005a). Dynamic Recrystallization Behavior in Martensite in 18Ni, 17Ni-0.2C and SM490 Steels. *Tetsu-to-Hagané* (Journal of the Iron and Steel Institute of Japan), Vol.91, pp. 602-608, ISSN 0021-1575.

Bao, Y.Z.; Adachi, Y.; Toomine, Y.; Xu, P.G.; Suzuki, T. & Tomota, Y. (2005b). Dynamic Recrystallization by Rapid Heating Followed by Compression for a 17Ni-0.2C Martensite Steel. *Scripta Materialia*, Vol.53, pp. 1471-1476, ISSN 1359-6462.

Bunge, H.J. (1982) *Texture Analysis in Materials Science*. Butterworth & Co., ISBN 0-408-10642-5, London, pp.88-98.

Chen, S.C.; Tomota, Y.; Shiota, Y.; Toomine, Y. & Kamiyama, T. (2006). Measurements of Volume Fraction and Carbon Concentration of the Retained Austenite by Neutron Diffraction. *Tetsu-to-Hagané* (Journal of the Iron and Steel Institute of Japan), Vol.92, pp.557-561, ISSN 0021-1575.

Doherty, R.D.; Hughes, D.A.; Humphreys, F.J.; Jonas, J.J.; Juul Jensen, D.; Kassner, M.E.; King, W.E.; McNelley, T.R.; McQueen H.J. & Rollett, A.D. (1997). Current issues in recrystallization: a review. *Materials Science and Engineering A*, Vol.238, pp.219-274, ISSN 0921-5093.

Dong, H. & Sun, X.J. (2006). Deformation induced ferrite transformation in low carbon steels. *Current Opinion in Solid State and Materials Science*, Vol.9, pp.269-276, ISSN 1359-0286.

Enomoto, M. & Furubayashi, E. (1977). A Crystallographic Study of Austenite Formation from Fe-Ni Martensite during Heating in Alpha-Gamma Region. *Transactions of the Japan Institute of Metals*, Vol.18, pp.817-824, ISSN 0021-4434.

Furuhara, T.; Yamaguchi, T.; Furimoto, S. & Maki, T. (2007). Formation of Ferrrite+Cementite microduplex structure by warm deformation in high carbon steels. *Material Science Forum*, Vol.539-543, pp.155-160, ISSN 0255-5476.

Glovre, G. & Selllers, C.M. (1973). Recovery and recrystallization during high temperature deformation of α-iron. *Metallurgical Transactions*, Vol.4, pp. 765-775, ISSN 0026-086X.

Hayashi, T.; Torizuka, S.; Mitsui, T.; Tsuzaki, K. & Nagai, K. (1999). Creation of low-carbon steel bars with fully fine ferrite grain structure through warm grooved rolling. *CAMP-ISIJ* (Current Advances in Materials and Processes), Vol.12, pp.385-388, ISSN 1882-8922.

Hayashi, T. & Nagai, K. (2002). Improvement of strength-ductility balance for low carbon ultrafine-grained steel through strain hardening design. *Transactions of the Japan Society of Mechanical Engineers. A*, Vol.68, pp.1553-1558, ISSN 0387-5008.

Hutchings, M.T.; Withers, P.J.; Holden, T.M. & Lorentzen, T. (2005). *Introduction to the Characterization of Residual Stress by Neutron Diffraction*, ISBN 0-415-31000-8, Taylor & Francis, New York, pp.230-238.

Larson, A.C. & Von Dreele, R.B. (2004). General Structure Analysis System (GSAS), *Los Alamos National Laboratory Report*. LAUR 86-748, pp.147-148.

Li, J.H.; Xu, P.G.; Tomota, Y. & Adachi, Y. (2008) Dynamic Recrystallization Behavior in a Low-carbon Martensite Steel by Warm Compression. *ISIJ International*, Vol.48, no.7, pp.1008-1013, ISSN 1485-1664.

Maki, T.; Okaguchi, S. & Tamura, I. (1982). Dynamic recrystallization in ferritic stainless steel. Strength of metals and alloys (ICSMA 6) : Proceedings of the 6th International Conference, pp.529-534., ISBN 0080293255. Melbourne, Australia, 16-20 August 1982.

Maki, T.; Furuhara, T.; & Tsuzaki, K. (2001). Microstructure Development by Thermomechanical Processing in Duplex Stainless Steel. *ISIJ International*, Vol..41. pp. 571-579. ISSN 1485-1664.

Miller, R.L. (1972). Ultrafine-grained microstructures and mechanical properties of alloy steels. *Metallurgical Transactions*, Vol.3, pp.905-912, ISSN 0026-086X.

Moriyama, M.; Takaki, S. & Kawagoishi, N. (2001). Influence of Reversion Austenite on Fatigue Property of 350 ksi Grade 18Ni Maraging Steel. *Journal of the Japan Society for Heat Treatment*, Vol.41. pp.266-271, ISSN 0288-0490.

Najafi-Zadeh, A.; Jonas, J.J. & Yue, S. (1992). Grain refinement by dynamic recrystallization during the simulated warm-rolling of interstitial free steels. *Metallurgical Transactions A*, Vol.23, pp.2607-2617. ISSN 0360-2133.

Ohmori, A. ; Torizuka, S. ; Nagai, K. ; Yamada K. & Kogo, Y. (2002). Evolution of ultrafine-grained structure through large strain-high Z deformation in a low carbon steel. *Tetsu-to-Hagané* (Journal of the Iron and Steel Institute of Japan), Vol.88, no. 12, pp.857-864, ISSN 0021-1575.

Ohmori, A.; Torizuka, S.; Nagai, K.; Yamada, K. & Kogo, Y. (2004). Effect of Deformation Temperature and Strain Rate on Evolution of Ultrafine Grained Structure through Single-Pass Large-Strain Warm Deformation in a Low Carbon Steel. *Materials Transactions*, Vol.45, pp.2224-2231, ISSN 1345-9678.

Poorganji, B.; Miyamoto, G.; Maki, T. & Furuhara, T. (2008). Formation of ultrafine grained ferrite by warm deformation of lath martensite in low-alloy steels with different carbon content. *Scripta Materialia*, Vol.59, pp.279-281, ISSN 1359-6462.

Reed, R.C. & Root, J.H. (1998). Determination of the Temperature Dependence of the Lattice Parameters of Cementite by Neutron Diffraction. *Scripta Materialia*, Vol.38, pp.95-99. ISSN 1359-6462.

Sawada, K.; Ohba, T.; Kushima, H. and Kimura, K. (2005). Effect of microstructure on elastic property at high temperatures in ferritic heat resistant steels. *Materials Science and Engineering A*, Vol.394, pp.36-42, ISSN 0921-5093.

te Velthuis, S.G.E.; Root, J.H.; Sietsma, J.; Rekveldt, M.T. & van der Zwaag, S. (1998). The ferrite and austenite lattice parameters of Fe-Co and Fe-Cu binary alloys as a function of temperature. *Acta Materialia*. Vol.46, pp.5223-5228, ISSN 1359-6454.

Tomota, Y.; Suzuki, T.; Kanie, A.; Shiota, Y.; Uno, M.; Moriai, A.; Minakawa, N. & Morii, Y. (2005). In situ neutron diffraction of heavily drawn steel wires with ultra-high strength under tensile loading. *Acta Materialia*. Vol.53, pp.463-467, ISSN 1359-6454.

Tomota, Y.; Narui, A. & Tsuchida, N. (2008). Tensile behavior of fine-grained steels, *ISIJ International*, Vol.48, pp.1107-1113, ISSN 1485-1664.

Torizuka, S. (2005) Production of ultrafine-grained steel bar and plate by high Z-large strain deformation in ferrite region. *Ferrum* (Bulletin of The Iron and Steel Institute of Japan), Vol.10, no.3, pp.188-195, ISSN 1341-688X.

Tsuji, N.; Matsubara, Y.; Saito, Y. & Maki, T. (1998). Occurance of Dynamic Recrystallization in ferritic Iron. *Journal of the Japan Institute of Metals*, Vol.62, pp.967-976. ISSN 0021-4876.

Tsuji, N.; Okuno, S.; Koizumi, Y. & Minamino, Y. (2004). Toughness of Ultrafine Grained Ferritic Steels Fabricated by ARB and Annealing Process. *Materials Transactions*, Vol. 45, no.7, pp.2272-2281, ISSN 1345-9678.

Ueji, R.; Tsuji, N.; Minamino, Y.; & Koizumi, Y. (2002). Ultragrain refinement of plain low carbon steel by cold-rolling and annealing of martensite. *Acta Materialia*, Vol.50, pp.4177-4189, ISSN 1359-6454.

Wenk, H.R.; Huensche, I. & Kestens, L. (2007). In-Situ Observation of Texture Changes during Phase Transformations in Ultra-Low-Carbon Steel. *Metallurgical and Materials Transactions A*, Vol.38, pp.261-267, ISSN 1073-5623.

Xu, P.G.; Tomota, Y.; Lukas, P.; Muransky, O. & Adachi, Y. (2006a). Austenite-to-ferrite transformation in low alloy steels during thermomechanically controlled process studied by in situ neutron diffraction. *Materials Science and Engineering A*, Vol.435, pp.46-53, ISSN 0921-5093.

Xu, P.G. and Tomota, Y. (2006b). Progress in materials characterization technique based on in situ neutron diffraction. *Acta Metallurgica Sinica*, Vol.42, pp.681-688, ISSN 0412-1961.

Xu, P.G.; Li, J.H.; Tomota, Y. and Adachi, Y. (2007). Effect of Carbon Addition on Ultrafine Grained Microstructure Formation by Warm Compression for Fe-18Ni Alloys, *Materials Science Forum*, Vol.558-559, pp.601-606, ISSN 0255-5476.

Xu, P.G.; Li, J.H.; Tomota, Y. & Adachi, Y. (2008a). Effects of Volume Fraction and Carbon Concentration of Austenite on Formation of Ultrafine Grained Ferrite/Austenite Duplex Microstructure by Warm Compression, *ISIJ International*, Vol.48, no.11, pp.1609-1617, ISSN 1485-1664.

Xu, P.G.; Tomota, Y. & Oliver, E.C. (2008b). Dynamic Recrystallization and Dynamic Precipitation Behaviors of a 17Ni-0.2C Martensite Steel Studied by *In Situ* Neutron Diffraction, *ISIJ International*, Vol.48, no.11, pp.1618-1625. ISSN 1485-1664.

Xu, P.G.; Tomota,Y.; Suzuki, T.; Yonemura, M. & Oliver, E.C. (2009). In Situ TOF Neutron Diffraction for Isothermal Ferrite Transformation during Thermomechanically Controlled Process of Low Alloy Steel. *Netsu Shori* (Journal of the Japan Society for Heat Treatment), Vol.49. special issue, pp.470-473. ISSN 0288-0490.

Xu, P.G.; Tomota, Y.; Vogel, S.C.; Suzuki, T.; Yonemura, M. & Kamiyama, T. (2012) Transformation Strain and Texture Evolution during Diffusional Phase Transformation of Low Alloy Steels Studied by Neutron Diffraction, *Reviews on Advanced Materials Science*. Vol.33. (in press). ISSN 1605-8127.

5

Ion-Beam-Induced Epitaxial Recrystallization Method and Its Recent Applications

Rossano Lang[1,4], Alan de Menezes[1], Adenilson dos Santos[2],
Shay Reboh[3,4], Eliermes Meneses[1], Livio Amaral[4] and Lisandro Cardoso[1]

[1]*Instituto de Física Gleb Wataghin - UNICAMP, Campinas, SP*
[2]*CCSST, Universidade Federal do Maranhão, Imperatriz, MA*
[3]*Groupe nMat, CEMES-CNRS, Toulouse*
[4]*Instituto de Física - UFRGS, Porto Alegre, RS*
[1,2,4]*Brazil*
[3]*France*

1. Introduction

The transition from amorphous Silicon to crystalline Silicon is a process of great technological importance and has raised an enormous interest also from a purely scientific perspective. Ion irradiation through an amorphous/crystalline interface may stimulate recrystallization or layer-by-layer amorphization depending on the sample temperature and ion beam parameters. In this chapter, we address some key features of this recrystallization phenomenon. The recrystallization/amorphization process will be discussed in relation to its dependence on the energy deposited during the ion beam irradiation, the sample temperature, and the presence of impurities, such as iron atoms dissolved within an amorphous silicon layer. Also it will be discussed the specific experimental condition under which a metastable phase of the $FeSi_2$ binary compound is trapped within a region of the recrystallized Silicon, in the form of nanoparticles. Theses nanoparticles, with different orientations and morphologies, are shown to cause interesting distortions in the surrounding Si crystal lattice. This effect reflects an important application of the ion-beam induced recrystallization process, as a method that could be used to synthesize ordered nanoparticles within a Silicon matrix.

2. Epitaxial crystallization

The amorphous Silicon (a-Si) is a Si phase with well-defined thermodynamic properties and that presents Gibbs free energy of ~ 0.12 eV/at. higher than that of the crystalline phase (c-Si) (Donovan et al., 1985, 1989; Roorda et al., 1989). This implies the existence of a driving force for the transition from the amorphous to the crystalline phase to occur. In other words, there is a natural tendency for the (a-Si) → (c-Si) transformation. At room temperature, the a-Si phase is metastable and it is transformed into c-Si only when submitted to high temperatures, typically higher than 450 °C (Olson & Roth, 1988; Williams, 1983). For the case of an amorphized layer on top of a Si single-crystal substrate, the transition occurs by a planar motion of the crystal-amorphous (c-a) interface, from the interior towards the surface, and hence decreasing the thickness of the existing amorphous layer with time, as schematically illustrated in Fig. 1. This effect is called epitaxial crystallization. For a pure thermal recrystallization process - SPEG (*Solid Phase Epitaxial Growth*), the growth rate is strongly dependent on the temperature and it presents an Arrhenius-like behavior with a unique activation energy of (2.68 ± 0.05) eV over a growth rate range of more than six orders of magnitude (Olson & Roth, 1988). For instance, at 470 °C the c-a interface displacement velocity is ~ 1 Å/min while at lower temperatures, this value decreases considerably and the amorphous to crystal transition becomes kinetically inhibited. However, epitaxial recrystallization of a-Si layers can also be achieved at lower temperatures (~ 200 - 320 °C) by ion-beam irradiation. This phenomenon represents a typical example of a dynamic annealing process and it is usually referred to as *Ion-Beam-Induced Epitaxial Crystallization* - IBIEC (Priolo & Rimini, 1990).

It is well known that both the implantation and the irradiation processes (depending on the dose and the beam ion mass) have the amorphization of the host matrix, as one of their main effects. For the specific case of Si, when the deposited energy (energy loss of ion beam mainly by nuclear collisions) handed by the projectile exceeds the threshold value of ~ 13 eV/at. a buried amorphous Si layer is formed (Narayan et al., 1984). However, if the Si is irradiated at a temperature above ~ 150 °C, there is a competition between both phenomena: amorphization and recrystallization.

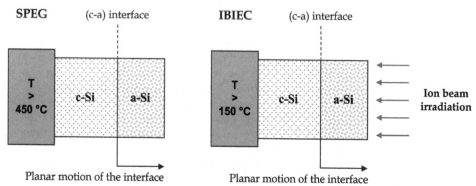

Fig. 1. Schematic illustration of two recrystallization processes in Silicon. SPEG (Solid Phase Epitaxial Growth) is a purely thermal process while, IBIEC is an ion beam induced epitaxial recrystallization process.

In fact, the temperature rise of the Si substrate is not solely responsible for this trend of the crystalline order recovery, since the recrystallization rate is greater than that obtained by a purely thermal process at the same temperature. A more complex mechanism involving a dynamic reordering stimulated by the ion beam is responsible for this effect. The recrystallization under ion-beam irradiation is the result of the prevalence of dynamic annealing over defect generation rate. Both processes depend on the substrate temperature and on the parameters of the irradiating beam, such as ion species, energy, ion dose and ion flux - also known as dose rate.

2.1 Crystal-amorphous interface displacement velocity

Rutherford backscattering spectrometry in combination with the ion channeling technique (RBS/C) is commonly used to directly monitor the c-a interface motion and infer the kinetics of the IBIEC process. The following RBS/C results were carried out with He+ beam at 1 MeV in a 170° scattering geometry. Figure 2 presents c-a interface position measurements of [100] oriented Si substrate previously implanted at room temperature with Fe+ ions at 100 keV energy. Channeled implantation was avoided by tilting the sample 7° normal with respect to the incident beam direction. Subsequently, the recrystallization of the Fe-implanted surface amorphous Si layer (filled circles in the figure) was induced by Si+ irradiation at high energy - 600 keV (whose projected range is well beyond the original c-a interface). In the IBIEC experiments the substrate temperature was fixed and controlled. This allows discrimination between the effects due to the heating of the sample holder and those due to ion-beam irradiation. In order to avoid beam heating effects low current density ($\approx$ 1 μA/cm²) was therefore used with ion flux of ~ 6.2x10¹² ions/cm²s. The substrate temperature was maintained at 350 °C. Essentially, one observes that the increase of the Si+ irradiated dose, leads to a decrease in the distance between the c-a interface and the surface. That is, there is a recrystallization towards the surface.

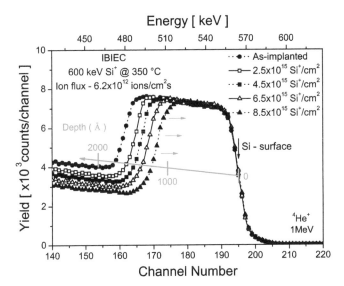

Fig. 2. Determination of the thickness of an amorphous layer using the ion channeling technique. This particular case corresponds to the recrystallization of ~ 176 nm a-Si under 600 keV Si+ beam irradiation keeping the substrate at 350 °C.

In implantation processes (Dearnaley et al., 1973), the implanted dose Φ is the integral of the ion flux in time, which corresponds to the total number of ions that were focused on the sample per unit area

$$\Phi \,(\text{ion/cm}^2) = \int_0^T \frac{I}{nq_e} dt \quad ; \quad T(\text{s}) = \frac{\Phi \cdot nq_e}{A \,/\, \text{cm}^2} , \tag{1}$$

where: I is the beam current in ampere per unit area of the sample (A/cm^2), q_e is the electron charge, $n = 1$ for once ionized ions, $n = 2$ for doubly ionized species, and so on. The term I/q_e is designated as the ion flux ϕ (ion/cm^2s) and T(s) is the implantation time in seconds. For the IBIEC, one designates a certain irradiation dose Φ and, consequently, the ion flux for such irradiation by $\phi = d\Phi/dt$. When the ion flux is kept constant at a given crystallization procedure, the irradiation dose may be interpreted as a measure of the processing time described by

$$\frac{dX_{c-a}}{dt} = \frac{dX_{c-a}}{d\Phi}\frac{d\Phi}{dt} = \phi\frac{dX_{c-a}}{d\Phi}.$$ (2)

Several reports define a c-a interface velocity R as the derivative of the recrystallized thickness (nm) *versus* dose (at/cm^2) curve. Thus, the unit of such "velocity" is expressed in nm/(at/cm^2). In the present study, figure 3 is an example that enables us to extract the interface displacement velocity or recrystallization rate.

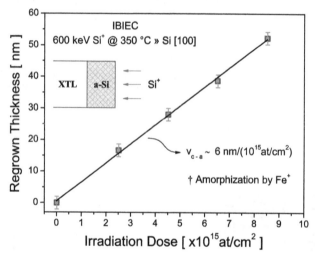

Fig. 3. Regrown thickness for [100] Si amorphized by Fe^+ ions and subsequently irradiated at 350 °C by 600 keV Si^+ ions. The interfacial displacement velocity R can be extracted from the curve.

It should be noted that a comparison between the thermal crystallization rate $V \equiv (dX_{c-a}/dt)$ and the recrystallization rate defined for the IBIEC $R \equiv (dX_{c-a}/d\Phi)$ is always possible, if one knows the ion flux ϕ of the process as shown in equation (2).

2.2 Influence of the beam parameters in the IBIEC process

Several experimental results have indicated that the beam parameters such as ion species (by the factor of nuclear energy loss S_n) and the ion flux ϕ have direct influence on the recrystallization rate. With regards to the energy loss by projectile-target elastic collisions, the following result is quite interesting. By using the recrystallized thickness *versus* irradiation dose plot (Fig. 3), the recrystallization rate R as a function of depth was determined as shown in figure 4. In the same figure, one observes the number of vacancies produced by irradiation of Si^+ ions as a function of depth. This evaluation corresponds to a calculation obtained by the SRIM algorithm with a displacement energy of 15 eV and a lattice binding energy of 2 eV) (Ziegler, 2011). The depth dependence of the regrowth experimental rates seems to follow the profile of defect generation.

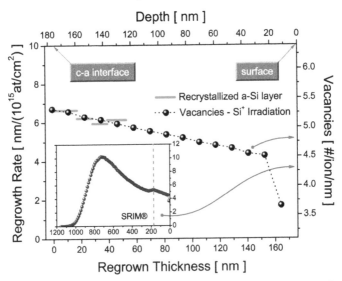

Fig. 4. Regrowth rate versus depth for a Si layer amorphized by Fe^+ ions and recrystallized at 350 °C by 600 keV Si^+ ions. The inset shows the number of vacancies generated by Si irradiation calculated by the SRIM code (closed circles).

This result suggests that the ion-induced recrystallization rate is associated with the production of point defects, or in a more general way, with the energy loss into elastic collisions at the c-a interface. Therefore, only those defects generated nearby or directly at the c-a interface are available for the recrystallization process. However, it should be mentioned that the diffusion or vacancy migration is not the promoter mechanism of the IBIEC. This issue will be addressed later. The abovementioned conclusions were also confirmed by other experimental results (Holmén et al., 1984; Linnros et al., 1984, 1985; Miyao et al., 1986; Williams et al., 1985). For instance, Linnros (Linnros et al., 1985) reported a clear experimental evidence of a linear dependence of the ion-induced recrystallization rate on the nuclear energy loss. In their experiments, ion beams of different masses (He, N, Ne, Si, As, Kr) were used to stimulate IBIEC where the recrystallization rate was observed to increase with increasing ion mass. This demonstrates that nuclear energy loss is the mechanism which produces the defects responsible for the IBIEC. Furthermore, the effects produced by both kinds of energy loss (nuclear and electronic) were discriminated through the dependence of the recrystallization rate on the beam energy. It was inferred that the electronic excitations and ionizations play practically no role in the recrystallization process (Elliman et al., 1985; Williams et al., 1985). In other experiments, in which a-Si layers were irradiated by the electron-beam, the recrystallization was observed only for energies above a threshold for atomic displacement of ~ 150 keV while below this threshold no epitaxial regrowth was produced, even after irradiation at very high electron doses (Lulli et al., 1987; Miyao et al., 1986; Washburn et al., 1983). Therefore, the observed epitaxy is associated with elastic collisions that transfer sufficient momentum to displace target atoms from their lattice site.

As mentioned earlier in this section, there is another parameter that directly influences the process - the ion flux. In general, it is observed that the lower the flux ϕ, the higher is the effective velocity $R \equiv (dX_{c-a}/d\Phi)$ defined by equation 2. This parameter is so important that high fluxes almost inhibit the process, especially for heavier ions (Linnros et al., 1985).

2.3 Temperature dependence

The sample temperature during irradiation is a fundamental variable in the IBIEC process (Elliman et al., 1985; Linnros et al., 1984; Priolo et al., 1988, 1989; Williams et al., 1985). Figure 5 shows, in an Arrhenius plot, the ion-induced growth rate (or recrystallization) as a function of the reciprocal temperature of an a-Si/Si(100) layer recrystallized by 600 keV Kr^{2+} ion irradiation (dose: 1x10^{15} Kr/cm^2 and ion flux: 1x10^{12} Kr/cm^{2}s). It also shows the recrystallization rate that represents the thermal contribution (SPEG) with an activation energy of (2.68 ± 0.05) eV. The data were extracted from references (Olson & Roth, 1988; Priolo et al., 1988) and reported in the figure. The growth rate is reported in Å/s (left-hand side) and in Å^4/eV (right-hand side). The latter scale represents the growth rate in the form $\Delta X/\Phi\nu(E)$, ΔX being the recrystallized thickness, Φ the dose and $\nu(E)$ the total energy deposited responsible for the displacement production at the c-a interface.

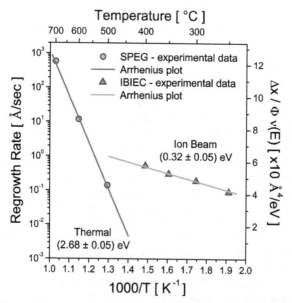

Fig. 5. Ion-induced growth rate versus reciprocal temperature for a-Si layers produced by Si$^+$ implantation and recrystallized by 600 keV Kr^{2+} ion irradiation. The thermal contribution to the growth rate is also displayed. Data extracted from references (Olson & Roth, 1988; Priolo et al., 1988) and reported in the figure.

It should be noted that recrystallization due to ion-beam irradiation occurs in a temperature range for which the thermal process is kinetically inhibited. Therefore, ion-beam irradiation strongly enhances the kinetics of recrystallization. For instance, at 250 °C the ion-induced growth rate is 0.07 Å/s while, an extrapolation of the thermal data gives a rate of only 10^{-10} Å/s. Furthermore, in the temperature range shown, the IBIEC presents an Arrhenius-like temperature dependence with an apparent activation energy of (0.32 ± 0.05) eV as also demonstrated by several other experiments (Elliman et al., 1985; Linnros et al., 1984; Williams et al., 1985).

It has been proposed that the activation energy for thermal recrystallization of a-Si layers is composed of two terms: one for defect generation and other for defect migration. Ion-beam irradiation clearly removes the main activated process usually limiting conventional thermal regrowth. In fact, during ion-beam irradiation defects are not thermally generated but rather being produced by means of atomic collisions. Jackson (Jackson, 1988) proposed that the activation energy inferred from the IBIEC is not associated with any activated process and therefore considered just as an apparent activation energy. The linear dependence of the regrowth rate as a function of reciprocal temperature, in a logarithmic plot, comes instead, from a balance between different effects. In addition, Jackson (Jackson, 1988) in his intracascade model suggested the dangling bond in the amorphous phase as the promoter for IBIEC. Dangling bonds are structural defects which by moving nearby the c-a interfacial region should produce a rearrangement of the bonds leading to recrystallization. In the mid-1990s, Priolo (Priolo et al., 1990) proposed a phenomenological model of IBIEC which combines the approach of the Jackson model (Jackson, 1988) with the structural and electronic features of models proposed for conventional thermal regrowth (Williams & Elliman, 1986). This model has explained all the experimental results so far.

2.4 Planar amorphization

In the previous sections, it was shown how a combination of thermal energy and energy deposited by ballistic effects can produce a non-equilibrium epitaxial recrystallization. However, IBIEC is a reversible process, where the increase in the ion flux of the irradiating beam, and /or the decrease in the target temperature can cause a planar layer-by-layer amorphization instead of an epitaxial recrystallization (Elliman et al., 1987; Linnros et al., 1986, 1988). Both processes are schematically illustrated in figure 6. At a constant flux, there is a substrate temperature T_R (reversal temperature) such that, when $T > T_R$, the ion irradiation produces epitaxial regrowth whereas, when $T < T_R$ the irradiation produces a layer-by-layer amorphization. The remarkable point is that the amorphization occurs just from the pre-existing amorphous seed and not from regions below the c-a interface, despite the fact that the energy loss by elastic collisions increases with the increasing depth. In this context, the net velocity of the interface motion can be described by two terms: a crystallization and an amorphization, as

$$R = \frac{dX_{c-a}}{d\Phi} = \frac{dX_{c-a}}{dt} / \phi = \left[\frac{dX_c}{dt} - \frac{dX_a}{dt} \right] / \phi . \tag{3}$$

When R is positive, the crystallization regime is prevalent and when R is negative the system is in the amorphization regime.

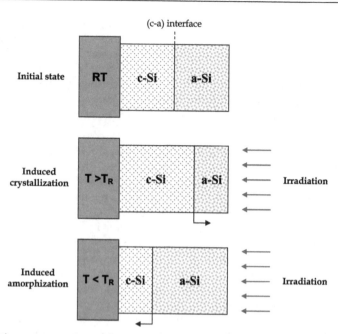

Fig. 6. Schematic representation of the ion-induced crystallization and amorphization. At temperatures below T_R the amorphous layer grows (amorphization regime), whereas at temperatures above T_R, it shrinks (crystallization regime).

An example of planar amorphization is shown in figure 7. In this example, a ~ 90 nm thick layer onto a Si(100) substrate was amorphized at room temperature by 40 keV Fe$^+$ ions at a dose of 1×10^{16} /cm^2. The irradiation was performed by using a 380 KeV Ne$^+$ beam at a dose of 1×10^{17} /cm^2 and ion flux of ~ 1.5×10^{13}/cm^{2}s. The substrate temperature was fixed at 100 °C. In the figure two RBS spectra are displayed in channeling condition for the a-Si layer before (open circles) and after (closed circles) Ne irradiation. A random spectrum is also reported. Ne irradiation clearly produces a great amount of damage beyond the original c-a interface. The a-Si surface layer is clearly seen to enlarge under a planar motion towards the sample interior. The amorphous layer has become ~ 50 nm thicker.

2.5 Impurity effects

Besides temperature and beam parameters, the kind of impurity dissolved in the sample plays an important role in the IBIEC. The presence of impurities dispersed within the a-Si layer can dramatically affect the recrystallization process. Depending on their behavior, they can be divided in two major categories: fast and slow diffusers. Fast diffusers comprehend species like Cu, Ag and Au which, at typical temperatures of the IBIEC process (~ 300 °C) have diffusivities of the order of 10^{-12} - 10^{-15} cm^2/s and low solid solubility. These impurities have therefore enough mobility to be redistributed at the advancing c-a interface during recrystallization, modifying the impurity initial profile through the segregation towards the surface imposed by the planar advance. On the other hand, the slow diffusers such as B, P, As, do not present the effect of segregation - are immobile in the time-temperature windows used during the IBIEC experiments, since the interface displacement velocity is much higher than the values of their mobilities in amorphous silicon. In this case, the initial concentration profile of these impurities remains unchanged after recrystallization. This allows one to produce non-equilibrium structures with impurities trapped in the c-Si at concentrations well above their solid solubility. However, the presence of impurities at the c-a interface can modify the recrystallization rate which is observed to increase or decrease, according to the particular species and to its concentration.

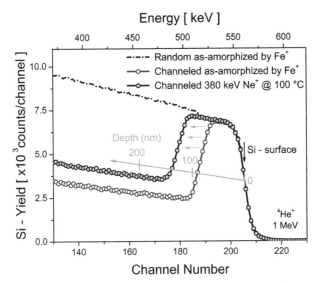

Fig. 7. RBS channeling spectra along the [100] Si axis for a ~ 90 nm thick amorphous layer before (open circles) and after (closed circles) 380 keV Ne^+ irradiation at 100 °C and dose of 1×10^{17} /cm². Note the planar growth of the amorphous phase.

From this brief discussion, it is noticed that the behavior of crucial parameters in the IBIEC process such as temperature and impurities is very similar to that in solid-phase epitaxial growth case. Despite the fact that the impact of these parameters is quite different for each phenomenon, there is a strong indication that similar microscopic processes occur in both cases. More precisely, the same interfacial defect responsible for the thermal recrystallization is considered to be active in IBIEC, being the ion beam the precursor of the increase in the average concentration of these defects.

3. Nanoparticles synthesis by IBIEC

The IBIEC technique has been used as a method to synthesize nanoparticles in Silicon matrix [Lang et al., 2010a, 2010b]. Specifically, the structural properties of the $FeSi_2$ nanoparticles synthesized in Fe^+ low dose implanted Si(100) substrates were investigated. Particularly in this experimental observation, the Fe proved to be a fast diffuser for IBIEC, despite of its action as a retardant of the process, whose recrystallization rate was dependent on the implanted Fe^+ concentration. Nevertheless, only the main results are reported here. The remarkable results which are presented show that the shape of the synthesized material (observed by high resolution transmission electron microscopy - HRTEM) singularly affects the surrounding Si lattice. The lattice strain shape-dependent distribution in both directions: out-of-plane and in-plane was tailored by X-ray Bragg-Surface Diffraction technique.

3.1 Fe$^+$ ion implantation in Si(100) and recrystallization process

A Si(100) n-type Czochralski wafer (thickness 500 μm, resistivity 10 - 20 Ω cm) was used as host matrix. Fe$^+$ ions at 40 keV were implanted at room temperature at an ion dose of $5x10^{15}$ cm^{-2}. Channeling effects were avoided by tilting the sample 7° normal with respect to the incident beam direction. The typical iron beam current density during implantation was about 150 nA/cm^2. Subsequently, ion-beam recrystallization experiments were performed at 350 °C using a 600 keV Si$^+$ beam (current density ≈ 1 μA/cm^2) to a total dose of $6x10^{16}$ ions/cm^2. The dose rate resulting from the Si beam current was $6.2x10^{12}$ ions/cm^{2}s. It is worth noting that Si$^+$ ions at 600 keV energy have a projected range (average depth) of ~ 770 nm with a straggle (standard deviation) of ~ 150 nm. These values ensure that the irradiation exceeds the pre-existing amorphous layer (well beyond the original c-a interface).

The structures obtained into the as-implanted and recrystallized samples were analyzed and characterized by Rutherford Backscattering Spectrometry combined with ion channeling technique (RBS/C - with He$^+$ beam at 1 MeV in a 170° scattering geometry) and also by transmission electron microscopy (TEM - JEOL 2010 operating at 200 kV). High-resolution rocking curves, as well as the reflection mappings of the Bragg-Surface Diffraction reflections, i.e., at the exact multiple diffraction condition, were carried out using the Huber multiaxis diffractometer of the XRD1 beam-line (Brazilian Synchrotron Radiation Facility - LNLS), with an incident beam wavelength of λ = 1.5495(5) Å, as defined by using a Si(111) channel-cut monochromator.

3.1.1 RBS and TEM – Results and discussion

Figure 8 shows RBS/C spectra obtained at random and [100]-channeled direction from the samples before (as-implanted) and after (recrystallized) irradiation. As observed in the aligned as-implanted spectrum, the implantation has produced an amorphous layer over ~ 90 nm while no channeling in the Fe signal was observed. However, expressive reduction of dechanneling yield in the Si profile is detected after IBIEC process (aligned IBIEC spectra relative to aligned as-implanted spectra). This decrease reflects the crystalline order recovery. The entire amorphous Si layer was recrystallized (at an average rate of ~ 0.04 nm/s) with a minimum backscattering yield on the subsurface region being χ_{min} ≈ 6.3 %. The RBS concentration-depth profile (not shown here) has indicated an implanted Fe peak concentration of ~ 2.8 at.% at 40 nm from the surface. However, the recrystallization process caused a slight narrowing of the Fe peak and a small segregation towards the surface. A significant degree of channeling (χ_{Fe} ≈ 46 %) was also observed in the Fe spectrum.

Figure 9 shows bright-field TEM cross-section images (taken at [110]$_{Si}$ zone axis) of the as-implanted sample. One clearly observes (Fig. 9a) the amorphous Si layer of ~ 90 nm produced by 40 keV Fe$^+$ implantation and the crystal-amorphous interface. Figure 9b shows in detail the c-a interface, where one can note a large amount of defects called "end-of-range defects" generated by the implantation process. This defective intermediate zone between the two distinct regions (crystalline and amorphous) is mainly composed of dangling bonds, and this particular kind of structural defect is responsible for the IBIEC process. Upon IBIEC conditions (temperature + irradiation) there is a dynamic rearrangement of these dangling bonds with annihilation in pairs which promotes the layer-by-layer planar recrystallization toward the surface.

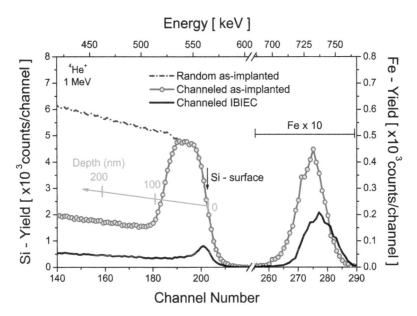

Fig. 8. 1 MeV He+ RBS spectra in channeling along the [100] direction from a 90 nm a-Si layer containing Fe before (as-implanted sample) and after (recrystallized sample) irradiation with 600 keV Si+ to a dose of 6×10^{16} ions/cm² at 350 °C.

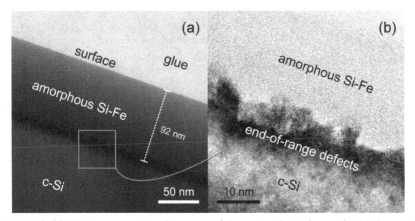

Fig. 9. Bright-field TEM cross-sectional images of the as-implanted sample. (a) Overview of the 90 nm thick a-Si layer containing Fe. (b) High-resolution of the selected region in (a) showing in detail the c-a interface region.

The complete recrystallization of the amorphous Si layer was also confirmed by TEM analyses. It should be noted that for a conventional thermal annealing at the same temperature (350 °C) and time (irradiation time ~ 160 minutes) would have produced a regrowth of only 7.7×10^{-3} nm which is a negligible amount. Therefore, ion-beam irradiation strongly enhances the kinetics of recrystallization. Figure 10 exhibits representative TEM micrographs of the recrystallized sample. The cross-section image, such as Fig. 10a taken along the $[110]_{Si}$ pole and slightly tilted on the zone axis, revealed the efficient a-Si regrowth and an impurity redistribution - nanoparticles formation after the IBIEC process. Iron was completely swept by the moving c-a interface and retained within the precipitated narrowing layer. Despite the amount of Fe present in the recrystallized region, the quality of the recovered crystal appears to be very good, as demonstrated by the TEM image. Three regions regarding the nanoparticles distribution are observed: a thin region a few nanometers thick which is closer to the surface (R_1); a Si region about 5 nm with a small concentration of nanoparticles (R_2); and a layer ($\approx$ 40 nm wide) with a higher concentration of nanoparticles (R_3).

High-resolution cross-sectional images (HRTEM) of the R_1 and R_3 regions are shown in the insets 10b, 10c and 10d. In inset 10b, it is possible to identify small irregular shaped nanoparticles at Si subsurface R_1. In the deeper layers (R_2 and R_3), two morphological variants of the metastable γ-FeSi$_2$ phase were observed and recognized: spherical-like nanoparticles epitaxially formed in the substrate with a fully aligned orientation regarding the Si matrix (Fig. 10c) and plate-like nanoparticles rotated with respect to the Si matrix (Fig. 10d) as previously reported (Lin et al., 1994). The spherical-like nanoparticles form coherent interfaces with the Si matrix, while the plate-like ones are elongated along $Si\langle\bar{1}12\rangle$ directions.

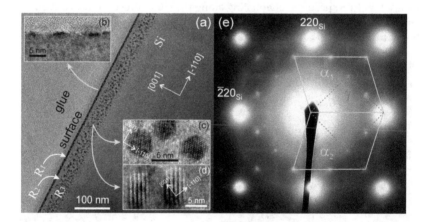

Fig. 10. TEM images of the recrystallized sample. **(a)** $[110]_{Si}$ cross-sectional revealing three nanoparticles regions in depth. HRTEM showing: **(b)** irregular shaped α-FeSi$_2$ nanoparticles at Si subsurface, **(c)** γ-FeSi$_2$ spherical-like and **(d)** γ-FeSi$_2$ plate-like nanoparticles in a deeper region. **(e)** $[001]_{Si}$ SAED pattern indicating the concomitant presence of both α- and γ-FeSi$_2$ phases.

Selected-area electron diffraction (SAED) pattern obtained from a plan-view specimen is presented in figure 10e. In addition to strong Bragg reflections of Si, extra spots due to the nanoparticles are apparent in the diffraction pattern. The extra spots show symmetric net patterns suggesting that there is a certain orientation relationship between the nanoparticles and the substrate lattice. From the analysis of the extra spots symmetry and lattice spacing, the diffraction pattern of Figure 10e can be explained as the overlap of cubic γ-FeSi$_2$ and tetragonal α-FeSi$_2$ phases. The reflections are consistent with $[\bar{1}12]_{\alpha 1}$, $[1\bar{1}\bar{2}]_{\alpha 2}$ (straight lines) and $[100]_\gamma$ (dotted line) net patterns (Behar et al., 1996; Vouroutzis et al., 2008). As the SAED measurements have identified two crystalline phases, the near-surface precipitated layer should contain α-FeSi$_2$ nanoparticles.

3.2 X-ray multiple diffraction

Bragg-Surface Diffraction (BSD) (Chang, 2004) is a special diffraction case of the X-ray multiple diffraction (XRMD) technique which has become a very useful and high resolution probe to study in-plane effects in single crystals in general, and also, with several interesting contributions to semiconductor epitaxial systems (dos Santos et al., 2009; Morelhão & Cardoso, 1993; Morelhão et al., 1998; Orloski et al., 2005; Lang et al., 2010b). For a more complete understanding of the experimental results that will follow in this chapter, we briefly discuss the physical aspects of the XRMD technique.

The XRMD phenomenon is systematically generated by aligning the primary planes of a single crystal - generally parallel to the sample surface, to diffract the incident beam and, by rotating it around the normal to the primary (h_p, k_p, l_p) planes while the diffracted beam is monitored by a detector. Under rotation (ϕ-axis), several other secondary (h_s, k_s, l_s) planes which are inclined with respect to the surface can enter into diffraction condition simultaneously with the primary ones. A closer observation of the diffraction geometry shows that other diffraction planes, the so-called coupling (h_c, k_c, l_c), also interact with the secondary diffracted beams to re-scatter them towards the detector. The obtained XRMD pattern, called Renninger scanning (RS) (Renninger, 1937), shows a series of peaks distributed according to the symmetry of the chosen primary vector and also to the symmetry plane established by rotation of the several reciprocal space secondary points when entering and leaving the Ewald sphere. Therefore, one can clearly observe in a RS, these two types of symmetry mirrors whose position and intensity distributions are essential for most of the applications of this technique. When a peak in the RS represents an interaction of the incident, the primary and one secondary beam, it shows up as a three-beam peak (or three-beam case). However, one can have two or three secondary beams simultaneously interacting to provide four or five-beam cases (or even cases for $n > 5$ interacting beams) being these secondary beams either Bragg (reflected) or Laue (transmitted) cases.

A special three-beam XRMD case, called Bragg-Surface Diffraction (BSD), appears under adequate conditions, that is, when the secondary diffracted beam propagates along the crystal surface under an extreme asymmetric geometry. A schematic diagram of the multiple scattering for the BSD case occurring inside the crystal, can be seen in Figure 11, where $\mathbf{H}_{ij}$ are the reciprocal lattice vectors corresponding to the primary planes ($\mathbf{H}_{01}$), secondary planes ($\mathbf{H}_{02}$) and coupling plane ($\mathbf{H}_{21}$). These BSD beams carry information on the sample surface which are useful for studying surface impurity incorporation effects (Lai et al., 2005), or even on the interface (layer/substrate), if such an interface is present, as in the semiconductor epitaxial structures. The technique has provided significant information when successfully applied as a special high-resolution 3D probe for the study of epitaxial layered heterostructures. Here, the layer and substrate lattices may be separately studied just by selecting one adequate layer or substrate peak. The technique is also a method capable of measuring strain fields at the interfaces in epilayer/substrate systems with depth penetration of 2 Å resolution and with enough sensitivity to detect lattices distortions in the range of 72 Å around the epilayer/substrate interface (Sun et al., 2006).

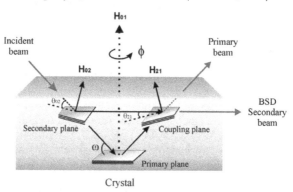

Fig. 11. BSD scheme using its consecutive scattering model with H_{01} (primary), H_{02} (secondary) and H_{21} (coupling) vectors. The coupling planes re-scatter the secondary beam towards the primary direction.

Besides the standard RS, another XRMD scanning methodology can give information on the crystalline quality obtained from an analysis of the $\omega{:}\phi$ mapping scans (Morelhão & Cardoso, 1996). By using this method, the ϕ rotation is performed for a range of ω angles each targeting an exact angular position of the multiple-beam Bragg condition. This approach results in a three dimensional plot of the primary intensity versus ω and ϕ in a coupled way from which, through analysis of the iso-intensity contours of such plots, one can obtain information on the lattice coherence along the beam path and hence, on the crystalline quality. It has been shown that when the FWHM (full width at half maximum) of the peak in the ϕ scan is larger in comparison to the one in the ω scan, there is almost no loss of coherence, i.e. confirming that the crystal is perfect or nearly perfect.

3.2.1 BSD reflections – Results and discussion

Measurements of the (004) symmetrical high resolution rocking curves (HRRC) are shown in figure 12 at two perpendicular orientations on the sample surface: $\phi = 0°$ (Fig. 12a) and 90° (Fig. 12b). Both patterns present practically the same result with two distinct peaks corresponding to R_2 and R_3 regions, clearly seen in each pattern, with smaller perpendicular lattice parameters in comparison to the matrix peak (stronger). Also, (002) HRRC were measured at ($\bar{1}11$) and (111) BSD reflections then, at two azimuth angles on the recrystallized sample ($\phi = -6.04°$ and 83.96°). The results are shown in figures 12c and 12d.

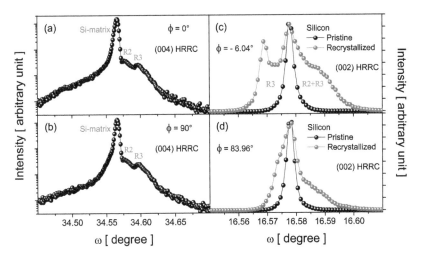

Fig. 12. High resolution rocking curves of the recrystallized sample for (004) reflection at $\phi = 0°$ (a) and $\phi = 90°$ (b) and (002) reflection at two BSD peaks: $-6.04°$ (c) and $83.96°$ (d). Si pristine is also added for comparison purposes.

As (002) is a forbidden reflection of the Si space group, no primary intensity can be observed out of the BSD secondary peaks. The rocking obtained at $\phi = -6.04°$ shows three different contributions: the stronger peak due to the Si matrix; the broad peak to the right (higher angles), due to the R_2 and R_3 convoluted peaks; and a distinct peak to the left (lower angles), probably associated only with the R_3 region. It should be noticed that the rocking at $\phi = 83.96°$ exhibits a meaningful profile difference, that is, the peak to the left (lower angles) does not appear as discriminated as in the previous measurement ($\phi = -6.04°$), it means, a noticeable confirmation of the anisotropic behavior. This anisotropy, observed between the $[\bar{1}10]$ and $[110]$ in-plane directions, could be associated with the plate-like nanoparticles (Fig. 10d) since the shape and orientation of these ordered nanoparticles should introduce different strains in both perpendicular directions.

Figure 13 shows the measured $\omega{:}\phi$ mappings for a Si matrix (pristine) and a recrystallized sample for comparison purposes to provide a better visualization and characterization of the detected anisotropy. The exact BSD reflection is tailored in both ω and ϕ-directions for each of the two above mentioned BSD secondary reflections: $(\bar{1}11)$ Si pristine in Fig. 13a and recrystallized sample in 13b and, for (111) Si pristine in 13c and recrystallized one in 13d. In fact, these mappings give a more complete 3D view of the BSD reflection condition which complements the 2D analysis obtained from the HRRC in Figs. 12c and 12d. Furthermore, the mappings allow for the lattice parameters and 2D-strain determination of the distorted regions (R_2 and R_3). The mappings obtained for the Si matrix (pristine) along the two in-plane perpendicular directions as depicted in 13a and 13c, exhibit only the BSD matrix peak at $\omega = 16.578°$ and, as expected, no difference is observed. In turn, the recrystallization process induces $FeSi_2$ nanoparticles nucleation within the implanted matrix and then, a huge broadening as well as a striking difference is clearly observed in Figs. 13b and 13d mappings. Besides the BSD matrix peak in Fig. 13b, two other peaks are also detected: one upper-side ($\omega \sim 16.59°$) and one lower-side ($\omega \sim 16.57°$) with respect to the matrix peak whereas, in Fig. 13d just the matrix and the upper-side peak are clearly seen since the lower-side peak appears as a shoulder of the matrix one. This result confirms the one obtained in Figs. 12c and 12d.

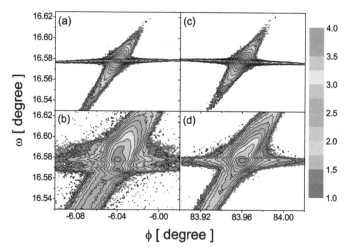

Fig. 13. Strain anisotropy in MBSD: Si pristine **(a)** and recrystallized sample **(b)** at ($\bar{1}11$) BSD ($\phi = -6.04°$) and Si pristine **(c)** and recrystallized **(d)** at (111) BSD ($\phi = 83.96°$).

Perpendicular and in-plane lattice parameters as well as strain values were obtained from the IBIEC sample $\omega{:}\phi$ mappings for $\phi = -6.04°$ ($\bar{1}11$) and $83.96°$ (111). The perpendicular strain obtained with the upper-side peaks of the BSD ($\bar{1}11$) and (111) $\omega{:}\phi$ mappings are $\varepsilon_\perp = -5.8(6)\times10^{-4}$ whereas, the in-plane strains are $\varepsilon_{||} = 0$. Then, as no anisotropy strain is detected from this upper-side peak ($\omega = 16.588°$) one can assume that most of this result can be assigned to the spherical-like nanoparticles rather than to the plate-like ones. In turn, for the lower-side peak, one observes that $\varepsilon_\perp$ and $\varepsilon_{||}$ values are distinct: $\varepsilon_\perp = 5.4(6)\times10^{-4}$ and $\varepsilon_{||} = 3.1(7)\times10^{-4}$ for ($\bar{1}11$) BSD peak and $\varepsilon_\perp = 2.4(8)\times10^{-4}$ and $\varepsilon_{||} = 1.3(9)\times10^{-4}$ for (111) and then, an analogous behavior happens for $a_\perp$ and $a_{||}$. Thus, one concludes there is anisotropy in the lattice parameters and strain in both sample directions: out-of-plane and in-plane. This anisotropy is attributed to the shape and distortion along the ($\bar{1}11$) and (111) crystallographic planes of the plate-like nanoparticles.

4. Conclusion

The conditions for the occurrence of the recrystallization and amorphization phenomena of a Si thin layer have been discussed in terms of the energy deposited by an ion beam as well as the sample temperature during an irradiation process. Also, the dependence of the recrystallization in relation to some impurity species, in particular, iron atoms dissolved into an amorphous Si layer has been discussed. Furthermore, it was shown how the ion-beam-induced epitaxial crystallization (IBIEC) process can be used as a method to synthesize nanoparticles within a Si matrix. For a specific case of Fe^+ low dose implanted in Si(100) substrate, nanoparticles with different orientations and morphologies were observed after the IBIEC process. These nanoparticles have caused interesting distortions in the surrounding Si lattice. In order to structurally characterize these distortions the Bragg Surface Diffraction was used as a 3D high-resolution tool. This non-conventional X-ray diffraction technique was able to discriminate between the out-of-plane and in-plane strain effects and to provide the direct observation of an important in-plane strain anisotropy.

5. References

Behar, M.; Bernas, H.; Desimoni, J.; Lin, X. W. & Maltez, R. L. (1996). Sequential phase formation by ion-induced epitaxy in Fe-implanted Si(001). Study of their properties and thermal behavior. *Journal of Applied Physics*, Vol. 79, No. 2, (January 1996), pp. 752-762, ISSN 1089-7550

Chang, S. L. (2004). *X-ray multiple-wave diffraction: theory and applications*, Springer Series in Solid-State Sciences, Vol. 143, Springer-Verlag, ISBN 3-540-21196-9, Berlin, Germany

Dearnaley, G.; Freeman, J. H.; Nelson, R. S. & Stephen, J. (1973). *Ion Implantation*, Noth-Holland Publishing Company, ISBN 0-7204-1758-9, Amsterdam, The Netherlands

Donovan, E. P.; Spaepen, F.; Turnbull, D.; Poate, J. M. & Jacobson, D. C. (1985). Calorimetric studies of crystallization and relaxation of amorphous Si and Ge prepared by ion implantation. *Journal of Applied Physics*, Vol. 57, No. 6, (March 1985), pp. 1795-1804, ISSN 1089-7550

Donovan, E. P.; Spaepen, F.; Poate, J. M. & Jacobson, D. C. (1989). Homogeneous and interfacial heat releases in amorphous Silicon. *Applied Physics Letters*, Vol. 55, No. 15, (October 1989), pp. 1516-1518, ISSN 1077-3118

dos Santos, A. O.; Lang, R.; de Menezes, A. S.; Meneses, E. A.; Amaral, L.; Reboh, S. & Cardoso, L. P. (2009). Synchrotron x-ray multiple diffraction in the study of Fe$^+$ ion implantation in Si(0 0 1). *Journal of Physics D: Applied Physics*, Vol. 42, No. 19, (October 2009), pp. 195401-195407, ISSN 1361-6463

Elliman, R. G.; William, J. S.; Maher, D. M. & Brown, W. L. (1985). Kinetics, microstructure and mechanisms of ion beam induced epitaxial crystallization of semiconductors. *Materials Research Society Symposium Proceedings*, Vol. 51, (January 1985), pp. 319-328, ISSN 0272-9172

Elliman, R. G.; Williams, J. S.; Brown, W. L.; Leiberich, A.; Maher, D. M. & Knoell, R. V. (1987). Ion-beam-induced crystallization and amorphization of silicon. *Nuclear Instruments and Methods in Physics Research Section B Beam Interactions with Materials and Atoms*, Vol. 19-20, No. Part 2, (January 1987), pp. 435-442, ISSN 0168-583X

Holmén, G.; Linnros, J. & Svensson, B. (1984). Influence of energy transfer in nuclear collisions on the ion beam annealing of amorphous layers in silicon. *Applied Physics Letters*, Vol. 45, No. 10, (November 1984), pp. 1116-1118, ISSN 1077-3118

Jackson, K. A. (1988). A defect model for ion-induced crystallization and amorphization. *Journal of Materials Research*, Vol. 3, No. 6, (November 1988), pp. 1218-1226, ISSN 2044-5326

Lai, X.; Roberts, K. J.; Bedzyk, M. J.; Lyman, P. F.; Cardoso, L. P. & Sasaki, J. M. (2005). Structure of habit-modifying trivalent transition metal cations (Mn^{3+}, Cr^{3+}) in nearly perfect single crystals of potassium dihydrogenphosphate as examined by X-ray standing waves, X-ray absorption spectroscopy, and molecular modeling. *Chemistry of Materials*, Vol. 17, No. 16, (August 2005), pp. 4053-4061, ISSN 1520-5002

Lang, R.; Amaral, L. & Meneses, E. A. (2010a). Indirect optical absorption and origin of the emission from β-FeSi$_2$ nanoparticles: Bound exciton (0.809 eV) and band to acceptor impurity (0.795 eV) transitions. *Journal of Applied Physics*, Vol. 107, No. 10, (May 2010), pp. 103508/1-103508/7, ISSN 1089-7550

Lang, R.; de Menezes, A. S.; dos Santos, A. O.; Reboh, S.; Meneses, E. A.; Amaral, L. & Cardoso, L. P. (2010b). X-ray Bragg-Surface Diffraction: a tool to study in-plane strain anisotropy due to ion-beam-induced epitaxial crystallization in Fe$^+$-implanted Si(001). *Crystal Growth and Design*, Vol. 10, No. 10, (August 2010), pp. 4363-4369, ISSN 1528-7505

Lin, X. W.; Washburn, J.; Liliental-Weber, Z. & Bernas, H. (1994). Coarsening and phase transition of FeSi$_2$ precipitates in Si. *Journal of Applied Physics*, Vol. 75, No. 9, (May 1994), pp. 4686-4694, ISSN 1089-7550

Linnros, J.; Svensson, B. & Holmén, G. (1984). Ion-beam-induced epitaxial regrowth of amorphous layers in silicon on sapphire. *Physical Review B*, Vol. 30, No. 7, (October 1984), pp. 3629-3638, ISSN 1550-235X

Linnros, J.; Holmén, G. & Svensson, B. (1985). Proportionality between ion-beam-induced epitaxial regrowth in silicon and nuclear energy deposition. *Physical Review B*, Vol. 32, No. 5, (September 1985), pp. 2770-2777, ISSN 1550-235X

Linnros, J.; Elliman, R. G. & Brown, W. L. (1986). The competition between ion beam induced epitaxial crystallization and amorphization in Silicon: The role of the divacancy. *Materials Research Society Symposium Proceedings*, Vol. 74, (January 1986), pp. 477-480, ISSN 0272-9172

Linnros, J.; Elliman, R. G. & Brown, W. L. (1988). Divacancy control of the balance between ion-beam-induced epitaxial cyrstallization and amorphization in silicon. *Journal of Materials Research*, Vol. 3, No. 6, (November 1988), pp. 1208-1211, ISSN 2044-5326.

Lulli, G.; Merli, P. G. & Antisari, M. V. (1987). Solid-phase epitaxy of amorphous silicon induced by electron irradiation at room temperature. *Physical Review B*, Vol. 36, No. 15, (November 1987), pp. 8038-8042, ISSN 1550-235X

Miyao, M.; Polman, A.; Sinke, W.; Saris, F. W. & van Kemp, R. (1986). Electron irradiation-activated low-temperature annealing of phosphorus-implanted silicon. *Applied Physics Letters*, Vol. 48, No. 17, (April 1986), pp. 1132-1134, ISSN 1077-3118

Morelhão, S. L. & Cardoso, L. P. (1993). Analysis of interfacial misfit dislocation by X-ray multiple diffraction. *Solid State Communications*, Vol. 88, No. 6, (November 1993), pp. 465-469, ISSN 0038-1098

Morelhão, S. L. & Cardoso, L. P. (1996). X-ray multiple diffraction phenomenon in the evaluation of semiconductor crystalline perfection. *Journal of Applied Crystallography*, Vol. 29, No. 4, (August 1996), pp. 446-456, ISSN 1600-5767

Morelhão, S. L.; Avanci, L. H.; Hayashi, M. A.; Cardoso, L. P. & Collins, S. P. (1998). Observation of coherent hybrid reflection with synchrotron radiation. *Applied Physics Letters*, Vol. 73, No. 15, (October 1998), pp. 2194-2196, ISSN 1077-3118

Narayan, J.; Fathy, D.; Oen, O. S. & Holland, O. W. (1984). High-resolution imaging of ion-implantation damage and mechanism of amortization in semiconductors. *Materials Letters*, Vol. 2, No. 3, (February 1984), pp. 211-218, ISSN 0167-577X

Olson, G. L. & Roth, J. A. (1988). Kinetics of solid phase crystallization in amorphous silicon. *Materials Science Reports*, Vol. 3, No. 1, (May 1988), pp. 1-77, ISSN: 0920-2307

Orloski, R. V.; Pudenzi, M. A. A.; Hayashi, M. A.; Swart, J. W. & Cardoso, L. P. (2005). X-ray multiple diffraction on the shallow junction of B in Si(0 0 1). *Journal of Molecular Catalysis A: Chemical*, Vol. 228, No. 1-2, (March 2005), pp. 177-182, ISSN 1381-1169

Priolo, F.; La Ferla, A. & Rimini, E. (1988). Ion-beam-assisted growth of doped Si layers. *Journal of Materials Research*, Vol. 3, No. 6, (November 1988), pp. 1212-1217, ISSN 2044-5326

Priolo, F.; Spinella, C.; La Ferla, A.; Rimini, E. & Ferla, G. (1989). Ion-assisted recrystallization of amorphous silicon. *Applied Surface Science*, Vol. 43, No. 1-4, (December 1989), pp. 178-186, ISSN 0169-4332

Priolo, F. & Rimini, E. (1990). Ion-beam-induced epitaxial crystallization and amorphization in silicon. *Materials Science Reports*, Vol. 5, No. 6, (June 1990), pp. 319-379, ISSN: 0920-2307

Priolo, F.; Spinella, C. & Rimini, E. (1990). Phenomenological description of ion-beam-induced epitaxial crystallization of amorphous silicon. *Physical Review B*, Vol. 41, No. 8, (March 1990), pp. 5235-5242, ISSN 1550-235X

Renninger, M. (1937). Umweganregung, eine bisher unbeachtete Wechselwirkungserscheinung bei Raumgitterinterferenzen. *Zeitschrift für Physik A Hadrons and Nuclei*, Vol. 106, No. 3-4, (July 1937), pp. 141-176

Roorda, S.; Doorn. S.; Sinke, W. C.; Scholte, P. M. L. O. & Van Loenen, E. (1989). Calorimetric Evidence for Structural Relaxation in Amorphous Silicon. *Physical Review Letters*, Vol. 62, No. 16, (April 1989), pp. 1880-1883, ISSN 1079-7114

Sun, W. C.; Chang, H. C.; Wu, B. K.; Chen, Y. R.; Chu, C. H.; Chang, S. L.; Hong, M.; Tang, M. T. & Stetsko, Y. P. (2006). Measuring interface strains at the atomic resolution in depth using x-ray Bragg-surface diffraction. *Applied Physics Letters*, Vol. 89, No. 9, (August 2006), pp. 091915/1-091915/3, ISSN 1077-3118

Vouroutzis, N.; Zorba, T. T.; Dimitriadis, C. A.; Paraskevopoulos, K. M.; Dózsa, L. & Molnár, G. (2008). Growth of β-FeSi$_2$ particles on silicon by reactive deposition epitaxy. *Journal of Alloys and Compounds*, Vol. 448, No. 1-2, (January 2008), pp. 202-205, ISSN 0925-8388

Washburn, J.; Murty, C. S.; Sadana, D.; Byrne, P.; Gronsky, R.; Cheung, N. & Kilaas, R. (1983). The crystalline to amorphous transformation in silicon. *Nuclear Instruments and Methods in Physics Research*, Vol. 209-210, No. Part 1, (May 1983), pp. 345-350, ISSN 0167-5087

Williams, J. S. (1983). Solid phase recrystallisation process in Silicon, In: Surface modification and alloying by laser, ion and electron beams, Poate, J. M.; Foti, G.; Jacobson, D. C., pp. 133, Plenum Press, ISBN 0306413736, New York, United States of America

Williams, J. S.; Elliman, R. G.; Brown, W. L. & Seidel, T. E. (1985). Dominant Influence of beam-induced interface rearrangement on solid-phase epitaxial crystallization of amorphous silicon. *Physical Review Letters*, Vol. 55, No. 14, (September 1985), pp. 1482-1485, ISSN 1079-7114

Williams, J. S. & Elliman, R. G. (1986). Role of electronic processes in epitaxial recrystallization of amorphous semicontluctors. *Physical Review Letters*, Vol. 51, No. 12, (September 1983), pp. 1069-1072, ISSN 1079-7114

Ziegler, J. F. (2011). *Stopping and range of ions in matter*, SRIM-2011. Available from http://www.srim.org/

6

The Deformability and Microstructural Aspects of Recrystallization Process in Hot-Deformed Fe-Ni Superalloy

Kazimierz J. Ducki
Silesian University of Technology
Poland

1. Introduction

The behaviour of metals and alloys during hot plastic working has a complex nature and it varies with the changing of such process parameters as (Zhou et al., 1994): deformation, strain rate and temperature. The high-temperature plastic deformation is coupled with dynamic recovery and recrystallization processes which influencing the structure and properties of alloys. One of crucial issues is finding the relationship between the hot plastic deformation process parameters, microstructure and properties. Since the sixties of the last century, theoretical and experimental investigations have been carried out to find those interdependence for steels and nickel alloys.

In the recent years, the constitutive equations describing hot plastic deformation processes have started to take into account the so-called internal variables determining the material condition. These variables include substructural parameters such as (Hansen, 1998; Sellars, 1998): grain size, grain shape, recrystalized volume fraction, dislocation density, subgrain size, subgrain misorientation angle and stacking fault energy (SFE). Determination of the above-mentioned parameters of a deformed material structure description requires the application of analytical methods primarily based on quantitative metallography and transmission electron microscopy (TEM). Taking those substructure parameters into consideration in calculations should enable the correct modelling of structural phenomena during hot plastic deformation and enhance the technological processes control for the purpose of obtaining the assumed structures of required properties (McQeen et al., 2002).

The Fe-Ni superalloys precipitation hardened by intermetallic phase of γ' - Ni$_3$(Al,Ti) type are one of the groups of construction materials intended for operation in cryogenic and elevated temperatures. These alloys are difficult to deform and are characterized by high values of yield stress at a high temperature. High deformation resistance of Fe-Ni alloys is caused by a complex phase composition, high activation energy of the hot plastic deformation process and a low rate of dynamic recrystallization. When choosing the conditions for hot plastic working of Fe-Ni alloys, the following factors should be considered (Bywater et al., 1976; Kohno et al., 1981; Ducki et al., 2006): the matrix grain size, plastic deformation parameters and the course of the recrystallization process. The grain size is of particular importance. Grain refining leads to an increased rate of recovery and dynamic recrystallization and to a smaller diameter of recrystallized grains. This is important, for the grain refinement in Fe-Ni superalloys has an advantageous influence on increasing their yield point and fatigue strength (Koul et al., 1994; Härkegård et al., 1998).

In the presented study, research has been undertaken on the influence of the initial microstructure of austenite and the parameters of hot plastic working on deformability, grain and subgrain size, and dislocation density in a high-temperature creep resisting Fe-Ni alloy. It is assumed that the results obtained will be used for optimizing hot plastic working processes and forecasting the microstructure and functional properties of products made of Fe-Ni superalloys.

2. Material and methodology

The examinations were performed on rolled bars, 16 mm in diameter, of an austenitic A-286 type alloy. The chemical composition is given in Tab. 1.

Content of an element [wt. %]															
C	Si	Mn	P	S	Cr	Ni	Mo	V	W	Ti	Al	Co	B	N	Fe
0.05	0.56	1.5	0.026	0.016	14.3	24.5	1.35	0.42	0.10	1.88	0.16	0.08	0.007	0.0062	55.3

Table 1. Chemical composition of the investigated Fe-Ni superalloy

In order to model the conditions of alloy heating prior to plastic processing, the investigations were carried out on samples after initial soaking at high temperatures. Sections of rolled bars, which the samples for investigations were made of, were subjected to two variants of preheating, i.e. 1100°C/2h and 1150°C/2h with subsequent cooling in water. Heat treatment of this type corresponds to the soaking parameters of the investigated superalloy before hot plastic processing (Kohno et al., 1981).

The research on the alloy deformability was performed in a hot torsion test on a Setaram torsion plastometer 7 MNG. The plastometric tests were performed every 50°C in a temperature range of 900÷1150°C, with a constant holding time of 10 minutes at the defined temperature. Solid cylindrical specimens (Ø 6.0 × 50 mm) were twisted at a rotational speed of 50 and 500 rpm, which corresponds to the strain rate of 0.1 and 1.0 s^{-1}, respectively. To freeze the structure, the specimens after deformation until failure were directly rapid cooled in water (Fig. 1).

The data obtained in the plastometric torsion test were entered in an Excel spreadsheet in the form of columns containing the recorded values. Processing of the measured data by means of filtration, cutting, shrinking and planishing was conducted using the Matlab 6 program. A correction of the torque moment, due to diversified rotational speed values and increase of the sample temperature during torsion, was calculated by the method of joint action of speed and temperature from the following relations (Hadasik, 2005):

$$M''' = M + \Delta M''' \tag{1}$$

$$\Delta M''' = M(N, \dot{N}, T) - M(N, \dot{N}_r, T + \Delta T) \tag{2}$$

$$\Delta M''' = A \cdot N^B \cdot \exp(C \cdot N) \cdot \dot{N}^{D+\frac{E}{T}} \cdot \exp\left(\frac{F}{T}\right) - A \cdot N^B \exp(C \cdot N) \cdot \dot{N}^{D+\frac{E}{T+\Delta T}} \cdot \exp\left(\frac{F}{T+\Delta T}\right) \quad (3)$$

where: M – recorded torque moment [Nm]; M''' – corrected torque moment value [Nm]; $\Delta M'''$ – torque moment correction taking account of a joint action of speed and temperature [Nm]; N – number of sample torsion, $\dot{N}$ – given torsion speed [rpm]; $\dot{N}_r$ – recorded rotational speed [rpm]; T – deformation temperature [°C]; ΔT – temperature increment during torsion [°C]; A, B, C, D, E, F – material constans.

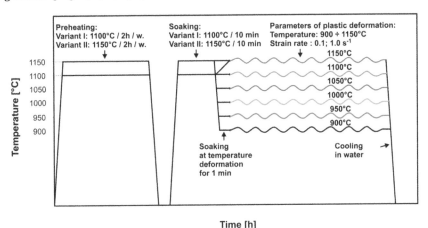

Fig. 1. Scheme of plastometric examination of the Fe-Ni alloy

The corrected data constituted a basis for the determination of equivalent deformation ε as a function of the number of the sample's rotations during torsion (Hadasik, 2005):

$$\varepsilon = \frac{2}{\sqrt{3}} \times \operatorname{arcsin} h\left(\frac{\pi \bar{R} N}{L}\right) \quad (4)$$

where: $\bar{R}$ equivalent radius corresponding to 2/3 of the outer radius R of the sample, L – measured sample length.

Yield stress σ_p was determined according to relation (5) taking account of the corrected torque moment M''', sample radius R, parameters m, p and axial force F_o (Hadasik, 2005):

$$\sigma_p = \left[\left(\frac{\sqrt{3} \cdot M'''}{2\pi R^3}\right)^2 \times (3+p+m)^2 + \left(\frac{F}{\pi R^2}\right)^2\right]^{0,5} \quad (5)$$

where: p – parameter reflecting stress sensitivity to deformation size; m – parameter reflecting stress sensitivity to deformation rate.

On the flow curves determined, the following parameters characterising plastic properties of the alloy in the torsion test were defined:

- σ_{pp} – maximum yield stress on the flow curve;
- ε_p – deformation corresponding to the maximum yield stress;
- σ_f – stress at which the sample is subject to failure;
- ε_f – deformation at which the sample is subject to failure, the so-called threshold deformation.

Relations between the yield stress and alloy deformation, and the deformation conditions were described using the Zener-Hollomon parameter Z (Zener et al., 1944):

$$Z = \dot{\varepsilon} \times \exp\left(\frac{Q}{RT}\right) = A \times \left[\sinh\left(\alpha\sigma_{pp}\right)\right]^{n} \tag{6}$$

where: $\dot{\varepsilon}$ – strain rate, Q – activation energy of the hot plastic deformation process, R – molar gas constant, T – temperature, and A, α, n – constants depending on grade of the investigated material.

The activation energy of the hot plastic deformation process Q was determined in accordance with the procedure specified in the work by (Schindler et al., 1998). The solution algorithm consisted in transforming equation (6) to the following form:

$$\dot{\varepsilon} = A \times \exp\left(\frac{-Q}{RT}\right)\left[\sinh(\alpha\sigma_{pp})\right]^{n} \tag{7}$$

Further procedure was based on solving equation (7) by a graphic method with the application of a regression analysis.

Structural inspections were performed on longitudinal microsections taken from the plastically deformed samples until failure after so-called "freezing" (Fig. 2). The specimens were etched using a reagent: 54 cm^3 HF, 8 cm^3 HNO$_3$ and 38 cm^3 distilled H$_2$O. Due to the deformation inhomogeneity, microscopic observation was conducted in a representative region located at a distance of ca. 0,65÷0,75 of the specimen radius.

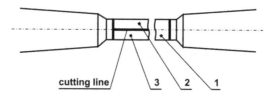

cutting line   3   2   1

1 - transverse microsection to axis sample
2 - longitudinal microsection to axis sample
3 - thin foil

Fig. 2. Scheme of material cutting for metallographic microsections and thin foils from plastometric samples

A quantitative analysis of the investigated structures was carried out by means of a computer program MET-ILO v. 3.0 (Szala, 1997). For the analyzed microstructures, in accordance with the methodology presented in paper (Cwajna et al., 1993), the following stereological parameters were determined:

- average area of grain plane section $\bar{A}$ [μm^2]:

$$\bar{A} = \frac{1}{N_A}$$ (8)

where: N_A – average grains number on area unit [μm^{-2}];

- variability coefficient of the grain plane section area $v(A)$:

$$v(A) = \frac{S(A)}{\bar{A}} \times 100 \, [\%]$$ (9)

where: $S(A)$ – empirical standard deviation of grain section area;

- volume fraction of dynamically recrystallized grains in the structure V_V [%];
- grain elongation coefficient δ (Feret coefficient):

$$\delta = \frac{F_x}{F_y}$$ (10)

where: $F_{x,y}$ – Feret diameters in x and y axes direction;

- classical, dimensionless shape coefficient ξ:

$$\xi = \frac{4\pi\bar{A}}{P^2}$$ (11)

where: P – perimeter of grain plane section.

The examination of the substructure was carried out by means of a JEM-100B Joel transmission electron microscope. Direct measurements on the TEM micrographs allowed the calculation of the structural parameters: the average subgrain area $\bar{A}$, and the mean dislocation density ρ. The mean subgrain areas were determined by a planimetric method making use of a semi-automatic image analyser MOP AMO 3 type. The measurements were conducted on the TEM images. The analysed microsections of thin foils involved measurements of about 150 subgrains for each sample. The mean dislocation density was calculated by use of a method based on counting the inter-section points of a network superimposed over the micrograph with dislocation lines. The dislocation density ρ as calculated for the thin foils according to the relation (Klaar et al., 1992):

$$\rho = \frac{x \cdot (n_1 / l_1 + n_2 / l_2)}{t} \quad [m^{-2}]$$ (12)

where: x – a coefficient which defines the fraction of invisible dislocations with Burgers vectors $a/2<111>$ for the A1 structure: $x = 2$ for image of dislocations observed in (111) reflex, $x = 1.5$ for image of dislocations observed in (200) reflex, $x = 1.5$ for image of dislocations observed in (220) reflex; $l_{1(2)}$ – the total length of the horizontal (vertical) lattice lines; $n_{1(2)}$ – the number of intersections of the horizontal (vertical) lattice lines with dislocations; t – the thickness of the foil.

The thickness of the foil in the investigated areas can be approximately calculated following the formula (Head et al., 1973):

$$t = n \bullet \zeta_{hkl} \tag{13}$$

where: n – a number of extinction lines; ζ_{hkl} – a value of extinction.

The values of ζ_{hkl} given by (Head et al., 1973) must be considered as a rough estimation of the actual value of extinction for the investigated material.

3. Results and discusion

3.1 Initial microstructure of the alloy

The application of two variants of initial solution heat treatment simulating the heating conditions allowed diversifying significantly the initial microstructure of the Fe-Ni alloy before hot plastic deformation. The alloy in its initial state differed primarily in the average grain size. After solution heat treatment under the conditions of 1100°C/2h/w., in the alloy microstructure presence was found of twinned austenite with medium-size grain ($\bar{A}$ = 2120 μm^2) with a small amount (ca. 0.3 wt.%) of insoluble particles (Fig. 3a). The increasing of the solution heat treatment temperature to 1150°C at an analogous soaking time resulted in an increase of the austenite grain ($\bar{A}$ = 6296 μm^2) and a reduction in the quantity and size of undissolved primary particles (Fig. 3b). For the analyzed variants of solution heat treatment, the microstructure of the samples was characterized by equiaxial grains, as evidenced by the elongation coefficient δ approximate to 1 and the dimensionless shape factor ζ of approximately 0.6.

Presence of titanium compounds, such as TiC carbide, $TiN_{0.3}$ nitride, $TiC_{0.3}N_{0.7}$ carbonitride, $Ti_4C_2S_2$ carbosulfide and Lavesa Ni_2Si phase and boride MoB was disclosed in the phase composition of the undissolved particles (Ducki, 2010).

a)   b)

Fig. 3. Diversified microstructure of alloy after solution heat treatment at: a) 1100°C/2 h/w., $\bar{A}$ = 2120 μm^2; b) 1150°C/2 h/w., $\bar{A}$ = 6296 μm^2

3.2 Deformability of the alloy

Knowledge of the phenomena occurring during hot deformation of materials enables selecting the correct conditions for plastic working and shaping their material characteristics. The results of the plastometric investigations, in the form of the calculated alloy flow curves at temperatures of 900÷1150°C for two options of initial soaking are shown in Fig. 4 and 5. The curves obtained for the option of initial soaking at 1100°C/2h and strain rate 0.1 s^{-1} have a shape characteristic of a material in which dynamic recovery and recrystallization take place (Fig. 4). High deformation values were obtained for the alloy in a wide range of torsion temperatures, i.e. 950÷1100°C. An increase of strain rate to 1.0 s^{-1} results in a significant increase of yield stress values and a distinct decrease of the alloy deformability at all temperatures analysed. This phenomenon can be explained by a higher alloy consolidation rate as well as too slow removal of the reinforcement as a result of dynamic recovery and recrystallization.

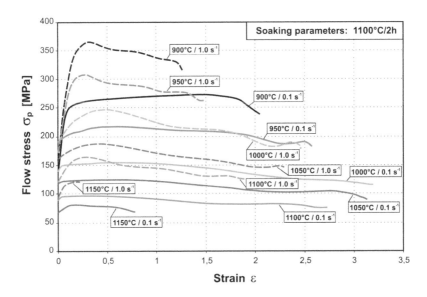

Fig. 4. The effect of deformation temperature on the flow stress of Fe-Ni alloy after initial soaking at 1100°C/2 h. Strain rate: 0.1 s^{-1} and 1.0 s^{-1}

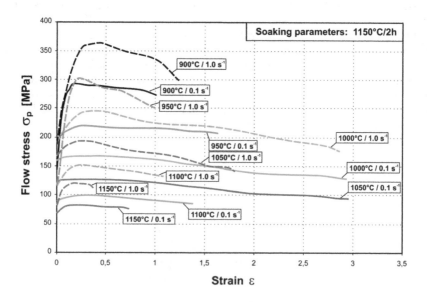

Fig. 5. The effect of deformation temperature on the flow stress of Fe-Ni alloy after initial soaking at 1150°C/2h. Strain rate: 0.1 s⁻¹ and 1.0 s⁻¹

An increase of the initial soaking temperature to 1150°C/2h significantly reduces the alloy deformability for the two strain rates, both at low and high deformation temperatures (Fig. 5). In this case, fairly high deformation values were obtained for the alloy in a narrow range of torsion temperatures, i.e. 1000÷1050°C. Such behavior of the alloy may be explained by a larger growth of austenite grains at this soaking temperature and, consequently, lower recovery and dynamic recrystallization rates.

The values determined for the maximum yield stress σ_{pp}, maximum deformation ε_p, stress until failure σ_f and threshold deformation ε_f depending on the temperature and strain rate are presented in Figs. 6-9. For the option of initial soaking at 1100°C/2h and torsion speed of 0.1 s⁻¹, the alloy under discussion shows a contiuous drop of σ_{pp} from values 277 MPa at a temperature of 900°C to the value of 81 MPa at 1150°C (Fig. 6). The threshold deformation ε_f rises initially together with the torsion temperature, reaching the maximum of (3.19/3.14) at 1000÷1050°C, and then falls (Fig. 8). An increase of the strain rate to 1.0 s⁻¹ results in an increase of σ_{pp} to maximum values of 367 MPa at the temperature of 900°C (Fig. 6) and a decrease of the threshold deformation to the maximum of 2.47/2.34 at 1000÷1050°C (Fig. 8).

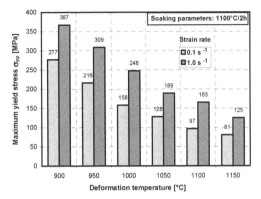

Fig. 6. The effect of deformation conditions on maximum flow stress. Initial alloy soaking: 1100°C/2 h

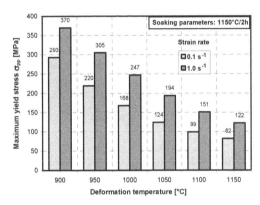

Fig. 7. The effect of deformation conditions on maximum flow stress. Initial alloy soaking: 1150°C/2 h

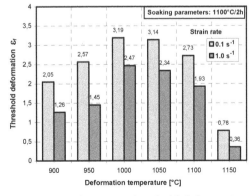

Fig. 8. The effect of deformation conditions on threshold deformation of the alloy. Initial soaking: 1100°C/2 h

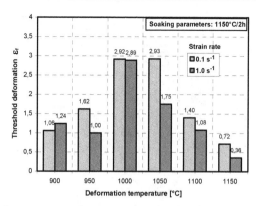

Fig. 9. The effect of deformation conditions on threshold deformation of the alloy. Initial soaking: 1150°C/2 h

An increase of the alloy initial soaking temperature to 1150°C/2h at a strain rate of 0.1 s⁻¹ results in a slight increase of σ_{pp} to maximum values of 293 MPa at 900°C (Fig. 7) and decrease of ε_f to the maximum of 2.92÷2.93 in the range of 1000÷1050°C (Fig. 9). An increase of the torsion speed to 1.0 s⁻¹ results in further increase of the σ_{pp} value to maximum values of 370 MPa at 900°C (Fig. 7), and decrease of ε_f to the maximum values of 2.89/1.75 at the temperature of 1000÷1050°C (Fig. 9).

The activation energy of the hot plastic deformation process Q was calculated by the means of a computer programme ENERGY 3.0 (Schindler et al., 1998). The activation energy, Q, necessary to initiate dynamic recrystallization in the Fe-Ni alloy, was determined on the basis of the linear dependencies presented in Fig. 10.

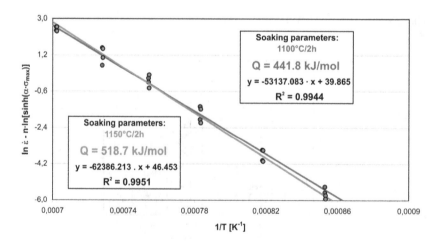

Fig. 10. The plot for determination of the activation energy for hot plastic deformation of the Fe-Ni alloy. Initial alloy soaking: 1100°C/2 h and 1150°C/2 h

The activation energy, Q, of hot plastic deformation for the Fe-Ni alloy depends on the temperature of initial soaking and equals as follows:

- $Q = 441.8$ [kJ/mol] – for initial alloy soaking 1100°C/2 h;
- $Q = 518.7$ [kJ/mol] – for initial alloy soaking 1150°C/2 h.

The higher value of the activation energy, Q, of hot plastic deformation for the alloy after initial soaking at 1150°C/2h can be justified by higher values of the maximum yield stress σ_{pp}, a larger growth of the initial austenite grain and a higher degree of matrix saturation with alloying elements.

The dependencies between maximum yield stress σ_{pp} and Zener Hollomon Z parameter are presented in Fig. 11. For both options of initial soaking, a power dependence ($R^2 = 0,98$) of the alloy yield stress was obtained as a function of the Z parameter. So determined function dependencies between the maximum yield stress σ_{pp} and the Z parameter had a form of power function:

- for the alloy after initial soaking 1100°C/2 h:

$$\sigma_{pp} = 0.43 \times Z^{0.151} \text{ [MPa]} \tag{14}$$

- for the alloy after initial soaking 1150°C/2 h:

$$\sigma_{pp} = 0.34 \times Z^{0.133} \text{ [MPa]} \tag{15}$$

Higher values of the Z parameter for the alloy after initial soaking at 1150°C/2h result from higher values of the plastic deformation activation energy Q.

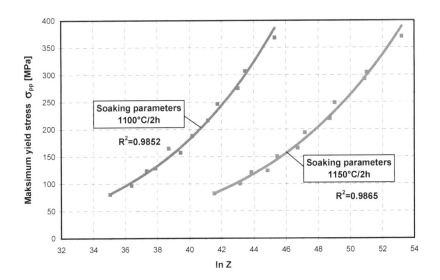

Fig. 11. Dependence of the maximum flow stress σ_{pp} on the Zener-Hollomon parameter Z. Initial alloy soaking: 1100°C/2 h and 1150°C/2 h

3.3 Microstructure of hot-deformed alloy

The recovery and dynamic recrystallization which occur in the Fe-Ni alloy during hot plastic deformation affect the size of the austenite grain. The results of investigations of the alloy microstructure after initial soaking at 1100°C/2h and deformation in a temperature range 900÷1150°C and a strain rate of 0.1 and 1.0 s⁻¹ are presented in Figs. 12a-d.

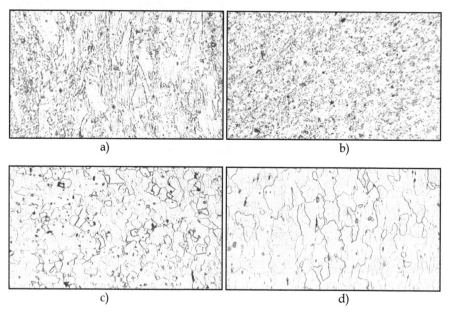

Fig. 12. The microstructure of the alloy after plastic deformation at: a) 900°C/1.0 s⁻¹, b) 950°C/0.1 s⁻¹, c) 1100°C/1.0 s⁻¹, d) 1150°C/0.1 s⁻¹. Initial soaking: 1100°C/2h

After deformation at 900°C for both of the torsion rates applied, the alloy structure is not completely recrystallized, which is indicated by the presence of primary elongated grains and fine recrystallized grains (Fig. 12a). At a torsion temperature in the range of 950÷1100°C, the alloy structure consisted of dynamically recrystallized grains (Fig. 12b and 12c). With an increasing deformation temperature, a gradual growth of the recrystallized grains was observed. The recrystallized grains were characterized by a deformed grain boundary line, which indicates an extensive cumulated deformation in the specimens. At the highest torsion temperature, 1150°C, some elongated grains of dynamically recrystallized austenite are observed in the alloy structure (Fig. 12d).

An increase of the initial soaking temperature of the alloy to 1150°C/2h results in increasing the initial austenite grain size and decreasing the kinetics of dynamic recovery and recrystallization for both of the analyzed strain rates (Figs. 13a-d). After deformation at 900°C/0.1 s⁻¹ and 900÷950°C/1.0 s⁻¹, the alloy microstructure was not completely recrystallized and it was composed of deformed primary grains and dynamically recrystallized grains of small sizes (Fig. 13a and 13b). The new, fine recrystallized grains nucleated at primary grain boundaries, creating the so-called "necklace". Within the deformation temperature range from 950÷1000 to 1100°C, the alloy structure is fine-grained and completely dynamically recrystallized (Fig. 13c). The highest deformation temperature of 1150°C induced deformation localization, as evidenced by the elongated dynamically recrystallized grains of varying sizes (Fig. 13d).

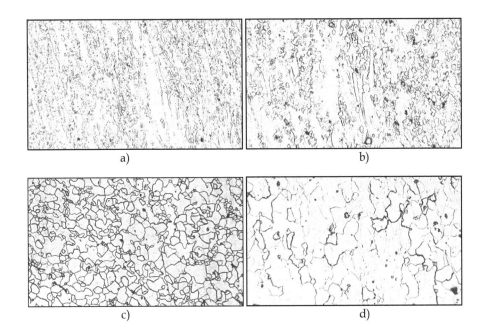

Fig. 13. The microstructure of the alloy after plastic deformation at: a) 900°C/1.0 s⁻¹, b) 950°C/0.1 s⁻¹, c) 1050°C/1.0 s⁻¹, d) 1150°C/0.1 s⁻¹. Initial soaking: 1150°C/2h

The results of a quantitative evaluation of the structure after initial soaking and deformation until failure in a temperature range of 900-1150°C and a strain rate of 0.1 and 1.0 s⁻¹ are presented in Figs. 14-17. In the structure of the alloy after initial soaking 1100°C/2h and deformation in the investigated temperature range at a strain rate 0.1 s⁻¹, monotonous growth of the grain average area $\bar{A}$ is observed from a value 16 μm² at 900°C to 198 μm² at 1150°C (Fig. 14). Up to the deformation temperature of 1100°C, the dynamically recrystallized grains are approximately equiaxial (δ = 0.99÷1.12), whereas at the highest torsion temperature, 1150°C, they are elongated (δ = 1.31) (Fig. 15). An increase of the strain rate to 1.0 s⁻¹ induces a certain reduction of the recrystallized grain size.

Also, in this case, in the investigated range of deformation temperatures, a monotonous growth of the grain average area $\bar{A}$ was observed, from 12 μm² at 900°C to 87 μm² at 1100°C (Fig. 14). In the analyzed range of torsion temperatures of 900÷1100°C, the dynamically recrystallized grains are approximately equiaxial (δ = 1.00÷1.09) (Fig. 15).

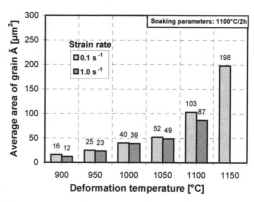

Fig. 14. The effect of deformation temperature on the average area of recrystallized grain after torsion at a rate of 0.1 and 1.0 s^{-1}. Initial soaking: 1100°C/2h

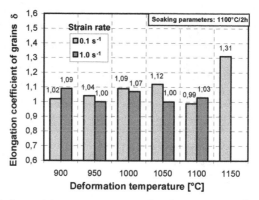

Fig. 15. The effect of deformation temperature on the elongation coefficient of recrystallized grain after torsion at a rate of 0.1 and 1.0 s^{-1}. Initial soaking: 1100°C/2h

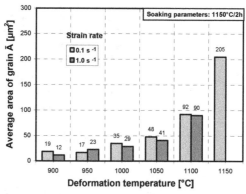

Fig. 16. The effect of deformation temperature on the average area of recrystallized grain after torsion at a rate of 0.1 and 1.0 s^{-1}. Initial soaking: 1150°C/2h

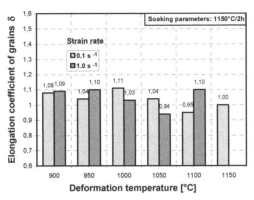

Fig. 17. The effect of deformation temperature on the elongation coefficient of recrystallized grain after torsion at a rate of 0.1 and 1.0 s^{-1}. Initial soaking: 1150°C/2h

An increase of the initial soaking temperature to 1150°C/2h and alloy deformation within the range of 900÷1150°C at a rate of 0.1 s^{-1} induces a similar reduction in size of the grain plane section area from 19 µm^2 at 900°C to 205 µm^2 at 1150°C (Fig. 16). An increase of the strain rate of samples to 1.0 s^{-1} within the deformation temperature range of 900÷1100°C causes a further reduction in the grain average area within the range from 12 to 90 µm^2. For both strain rates, the grains after dynamic recrystallization are approximately equiaxial ($\delta = 0.94÷1.11$) (Fig. 17).

A comparison of the size of recrystallized grain in the Fe-Ni alloy after deformation at a strain rate of 0.1 and 1.0 s^{-1} for two variants of initial soaking at 1100°C/2 h and 1150°C/2 h is presented in Fig. 18 and 19. After deformation at a strain rate of 0.1 s^{-1} within the temperature range of 900÷1150°C for both variants of initial soaking, a similar grain size $\bar{A}$ was obtained in the range from 16 to 205 µm^2 (Fig. 18). A higher strain rate of 1.0 s^{-1} within the temperature range of 900÷1100°C allows obtaining a slightly higher degree of grain refining in the range from 12 to 90 µm^2 (Fig. 19). Thus, it can be concluded that the initial microstructure of an alloy after initial soaking has no significant influence on the size of recrystallized grain after plastic deformation.

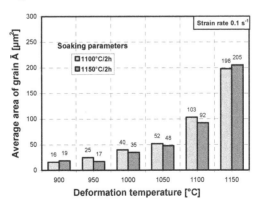

Fig. 18. The effect of deformation temperature on the average area of recrystallized grain after initial soaking of the alloy at 1100°C/2 h and 1150°C/2 h

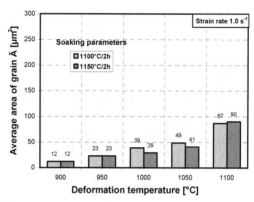

Fig. 19. The effect of deformation temperature on the average area of recrystallized grain after initial soaking of the alloy at 1100°C/2 h and 1150°C/2 h

The average grain plane section area of samples deformed under the same conditions but with a different initial grain size after initial soaking is similar. The average size of recrystallized austenite grain depends mainly on the deformation temperature and, to a lesser degree, on the strain rate applied for the alloy (Fig. 20 and 21).

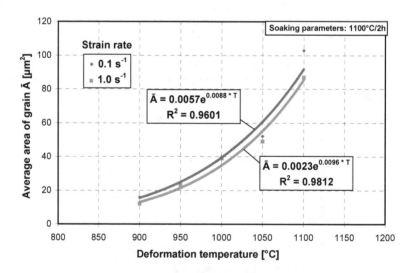

Fig. 20. Relationship between the average grain area after recrystallization versus deformation temperature and strain rate. Initial alloy soaking: 1100°C/2 h

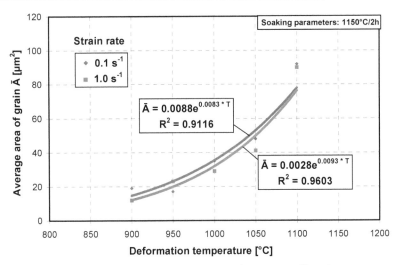

Fig. 21. Relationship between the average grain area after recrystallization versus deformation temperature and strain rate. Initial alloy soaking: 1150°C/2 h

An assessment of the degree of influence of the deformation temperature and strain rate on the average grain plane section area of the recrystallized Fe-Ni alloy was obtained after introducing the Zener-Hollomon Z parameter. As appears from the dependencies developed, the fine-grained microstructure of the alloy after deformation was obtained more easily after initial soaking at a temperature of 1100°C/2h when compared to soaking at 1150°C/2 h (Fig. 22). This is evidenced by lower Z parameter values for the deformation after initial soaking at 1100°C/2 h: a low strain rate and a moderate deformation temperature.

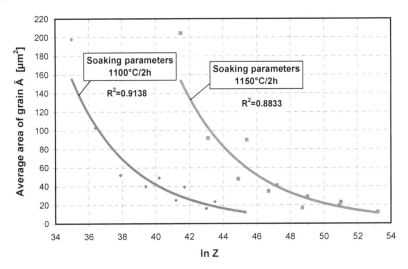

Fig. 22. Relationship between the Zener-Hollomon parameter and the average grain area after recrystallization. Initial alloy soaking: 1100°C/2 h and 1150°C/2 h

For both variants of initial soaking of the alloy, relationships were determined between the average plane section area of the recrystallized grain $\bar{A}$ and the Z parameter (eq. 16 and 17):

- for the alloy after initial soaking 1100°C/2h:

$$\bar{A} = 9.3 \times 10^5 \times Z^{-0.250} \; [\mu m^2] \tag{16}$$

- for the alloy after initial soaking 1150°C/2h:

$$\bar{A} = 2.0 \times 10^6 \times Z^{-0.224} \; [\mu m^2] \tag{17}$$

The determined relationships are essential where the critical value of a significance test for the direction factor of regression line "p" is less than 0.05. In the analyzed case, the "p" factor value for both variants of initial soaking of the alloy, i.e. at 1100°C/2h and 1150°C/2h, equaled 3.08×10^{-5} and 5.03×10^{-5}, respectively, which indicates the significance of the determined relationships.

3.4 Substructure of hot-deformed alloy

The recovery and dynamic recrystallization processes during hot plastic deformation of the Fe-Ni alloy cause changes in the dislocation and subgrain substructures. An analysis of the microscopic examination results allows affirming that the nature and extent of changes in the alloy substructure was dependent on the deformation temperature and strain rate, and the conditions of initial soaking. After initial soaking (1100°C/2 h) and deformation at a temperature of 900°C and a rate of 0.1 and 1.0 s⁻¹, effects of dynamic recovery and dynamic recrystallization (Fig. 23 and 24) were observed in the alloy microstructure. In grain areas with major defects of the austenite, a cellular dislocation substructure and subgrains with different dislocation densities were formed.

Fig. 23. Alloy microstructure after soaking at 1100°C/2 h and deformation at 900°C/0.1 s⁻¹. Subgrains and recrystallized grains

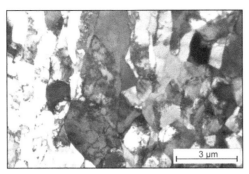

Fig. 24. Alloy microstructure after soaking at 1100°C/2 h and deformation at 900°C/1.0 s⁻¹. Subgrain structure formation

At a higher deformation temperature, e.g. 950°C, the proceeding recovery processes were accompanied by intensive dynamic recrystallization (Fig. 25 and 26). In the alloy substructure, dynamically recrystallized grains were observed next to the subgrains (Fig. 25). In the samples deformed at a higher rate, 1.0 s⁻¹, at this temperature the fraction and size of dynamically recrystallized microregions increase (Fig. 26).

The alloy deformed within the temperature range of 1000÷1050°C is characterized by a microstructure typical of a dynamically recrystallized material (Fig. 27 and 28). The austenite microstructure is composed predominantly of recrystallized grains free of dislocations (Fig. 27). Further perfecting of the substructure is observed in the neighbouring subgrains, as evidenced by equiaxiality of the subgrains and the decreasing density of dislocations inside them (Fig. 28). It was found that a higher strain rate (1.0 s⁻¹) leads to a growth of the subgrain and a reduction of the dislocation density.

Fig. 25. Alloy microstructure after soaking at 1100°C/2 h and deformation at 950°C/0.1 s⁻¹. Subgrains and recrystallized grains

Fig. 26. Alloy microstructure after soaking at 1100°C/2 h and deformation at 950°C/1.0 s⁻¹.
Regions of recrystallized austenite and subgrains

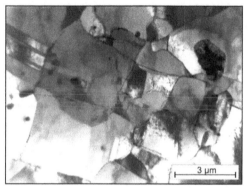

Fig. 27. Alloy microstructure after soaking at 1100°C/2 h and deformation at 1000°C/0.1 s⁻¹.
Recrystallized grains with twins and subgrains

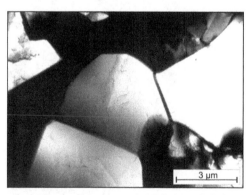

Fig. 28. Alloy microstructure after soaking at 1100°C/2 h and deformation at 1050°C/1.0 s⁻¹.
The process of further perfecting of the subgrain structure

In the alloy substructure, after deformation at the highest temperature of 1100÷1150°C, a reincrease of the dislocation density and repolygonization were observed (Fig. 29). At a higher strain rate of 1.0 s⁻¹, areas with recrystallized grain free of dislocations are dominant in the alloy substructure (Fig. 30).

An increase of the initial soaking temperature to 1150°C/2h inhibits the dynamic recovery and recrystallization processes (Fig. 31 and 32). In the regions with austenite subgrain after deformation at a temperature of 900°C at a rate of 0.1 s⁻¹, deformation microtwins appear (Fig. 31). The subgrains being formed, especially at a high strain rate of 1.0 s⁻¹, have an elongated shape and different dislocation densities (Fig. 32).

At a higher deformation temperature, e.g. 1000°C, and a strain rate of 0.1 and 1.0 s⁻¹, the alloy substructure reconstruction is inhomogeneous. It is characterized by the presence of areas where structural changes are inhibited and accelerated (Fig. 33 and 34).

Effects of dynamic recovery (Fig. 33) and dynamic recrystallization (Fig. 34) were found there. The growth of new grains in the dynamic recrystallization process proceeded through the coalescence of subgrains and their subsequent growth (Fig. 35). The bending of the grain boundary towards areas with a higher dislocation density indicates the direction of the boundary movement. The principal mechanism of the coalescence includes reactions between dislocations which lead to disappearance of the dislocation boundary and formation of grain from the combination of several neighboring subgrains.

The dislocation density in the subgrain area does not decrease significantly when increasing the deformation temperature to 1100°C compared to a lower deformation temperature (Fig. 36 and 37). The dynamic deformation phenomena are accompanied by a continuous process of structural reconstruction of the material, i.e. repolygonization. It consists in re-saturation of subgrains with dislocations and their rearrangement together with the creation of new subboundaries and walls of a polygonal type (Fig. 36).

Deformation of the alloy at a higher rate of 1.0 s⁻¹ was accompanied in the substructure by dislocation rearrangement with the formation of polygonal walls and a cellular substructure (Fig. 37). The phenomenon of repolygonization in the austenite grains was observed at the highest deformation temperature (1100÷1150°C).

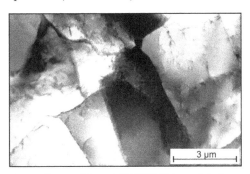

Fig. 29. Alloy microstructure after soaking at 1100°C/2 h and deformation at 1100°C/0.1 s⁻¹. The effects of repolygonization in austenite subgrains

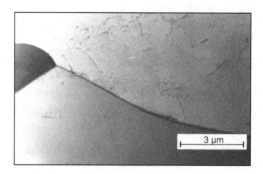

Fig. 30. Alloy microstructure after soaking at 1100°C/2 h and deformation at 1100°C/1.0 s⁻¹. Large austenite subgrains with a low dislocation density

Fig. 31. Alloy microstructure after soaking at 1150°C/2 h and deformation at 900°C/0.1 s⁻¹. Subgrain structure and grains with microtwins

Fig. 32. Alloy microstructure after soaking at 1150°C/2 h and deformation at 900°C/1.0 s⁻¹. Non-equiaxed, austenite subgrains

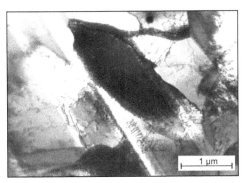

Fig. 33. Alloy microstructure after soaking at 1150°C/2 h and deformation at 1000°C/0.1 s^{-1}. The effects of polygonization and deformation microtwins

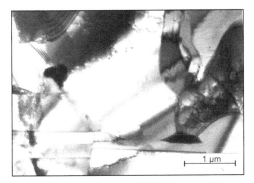

Fig. 34. Alloy microstructure after soaking at 1150°C/2 h and deformation at 1000°C/1.0 s^{-1}. The grains after dynamic recrystallization

Fig. 35. Alloy microstructure after soaking at 1150°C/2 h and deformation at 1000°C/0.1 s^{-1}. The recrystallized grain formation as a result of coalescence of subgrains

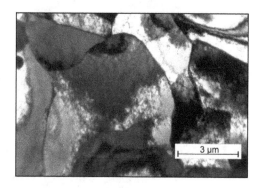

Fig. 36. Alloy microstructure after soaking at 1150°C/2 h and deformation at 1100°C/0.1 s⁻¹. The formed subgrain structure with dislocations

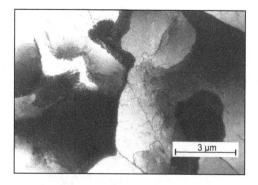

Fig. 37. Alloy microstructure after soaking at 1150°C/2 h and deformation at 1100°C/1.0 s⁻¹. Austenite repolygonization and a cellular dislocation structure

The course of changes in the subgrain sizes depending on the temperature deformation and strain rate for the two variants of initial soaking of the alloy is shown in Fig. 38 and 39. It was found that an increase in the alloy deformation temperature from 900 to 1150°C results in a growth of the subgrain.

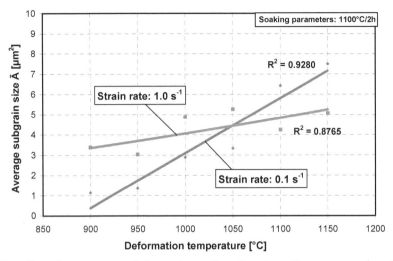

Fig. 38. The effect of temperature deformation and strain rate on the average subgrain size. Initial alloy soaking: 1100°C/2 h

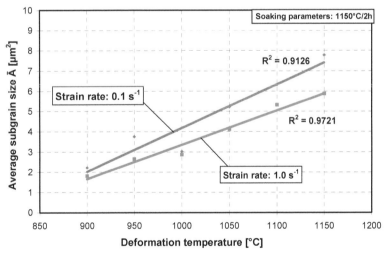

Fig. 39. The effect of temperature deformation and strain rate on the average subgrain size. Initial alloy soaking: 1150°C/2 h

However, no influence was observed of the conditions of initial soaking on the subgrain size. The average area of the subgrain plane section $\bar{A}$ varied within the range from 1.0 μm^2 to 7.8 μm^2 for both variants of initial soaking. The influence of the strain rate on the subgrain size was more significant, in particular for the initial soaking at 1100°C/2 h (Fig. 38). More intensive changes in the subgrain size were observed at a low strain rate of 0.1 s^{-1}, which can be explained by a higher cumulative deformation in the samples.

The dislocation density depending on the deformation temperature and strain rate for the two variants of initial soaking of the alloy is shown in Fig. 40 and 41. An increase of the deformation temperature was accompanied by a decreasing dislocation density. No significant influence was found of the initial soaking parameters on the dislocation density.

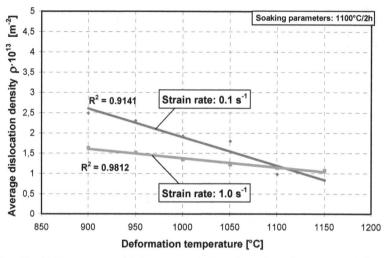

Fig. 40. The effect of temperature deformation and strain rate on the average dislocation density. Initial alloy soaking: 1100°C/2 h

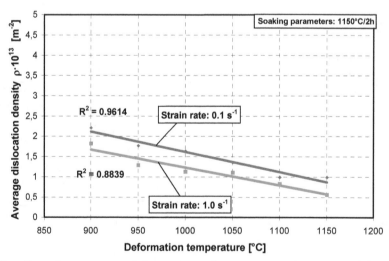

Fig. 41. The effect of temperature deformation and strain rate on the average dislocation density. Initial alloy soaking: 1150°C/2 h

For both variants of initial soaking, the dislocation density varied within a narrow range from 0.8×10^{13} m^{-2} to 2.5×10^{13} m^{-2}. The gradual reduction in the dislocation density observed in the samples as the deformation temperature increased from 900 to 1150°C shows a continuous process of substructure reconstruction and redeformation. For both variants of initial soaking of the alloy, higher dislocation densities were obtained for the lower strain rate (0.1 s⁻¹), which can be explained by a higher cumulative deformation in the material.

4. Summary

The work analyzes the relationships between the conditions of hot plastic deformation and deformability and microstructure of an austenitic Fe-Ni superalloy precipitation-strengthened by phase γ' type. The hot torsion tests carried out in the range of temperature of 900÷1150°C, at a strain rate of 0.1 s⁻¹ and 1.0 s⁻¹ allowed determining the influence of the initial soaking conditions and deformation parameters on technological plasticity of the Fe-Ni superalloy, as well as on the strengthening and structure reconstruction processes. It was found that optimal values of the maximum yield stress σ_{pp} and threshold deformation ε_f, as well as the required fine-grain microstructure, were obtained for the alloy after initial soaking at 1100°C/2 h and deformation at a rate of 0.1 s⁻¹ in the temperature range of 1050÷950°C. The increase of yield stress, σ_{pp}, and the decrease of deformability of the alloy, ε_f, as the initial soaking temperature was rising up to 1150°C/2 h, with the deformation rate increasing to 1.0 s⁻¹, was associated with a growth of the initial grain size and the degree of austenite saturation with alloying elements. As a result of these processes, the stacking fault energy (SFE) of the austenite increased and so did the ability of the material to strengthen. This, in turn, led to an increase of activation energy of the hot plastic deformation process from the value $Q = 441.8$ kJ/mol (after initial soaking at 1100°C/2 h) to $Q = 518.7$ kJ/mol (after initial soaking at 1150°C/2 h).

An analysis of the flow curves and the examination results of the Fe-Ni alloy microstructure and substructure revealed dynamic recovery, recrystallization and repolygonization, occurring consecutively in the course of hot deformation. None of the detected stages of changes in the alloy structure constituted an independent process. Their course depended on both, the deformation parameters $(T, \dot{\varepsilon})$ and the initial soaking conditions. The growth of new grains in the dynamic recrystallization process took place through coalescence of subgrains and their subsequent growth. For both variants of initial soaking of the alloy, the analyzed quantitative indicators of the substructure depended fundamentally on the deformation temperature and, to a lesser degree, on the strain rate. The average size of subgrains $\bar{A}$ increased from 1.0 µm² to 7.8 µm² as the deformation temperature rose from 900°C to 1150°C. The average dislocation density ρ decreased gradually in the range from 2.5×10^{13} m⁻² to 0.8×10^{13} m⁻² as the deformation temperature rose in the range of 900÷1150°C.

The dynamic recrystallization proceeding in the Fe-Ni alloy during hot plastic deformation caused high refinement of the material structure. The average area $\bar{A}$ of recrystallized grains increased as the deformation temperature rose and it changed in the analyzed range of deformation parameters from 16 µm² to 205 µm², which, with reference to the initial grain size, meant refinement of the alloy structure of the order of 10÷370×. The average size of recrystallized austenite grain depended mainly on the deformation temperature and, to a lesser degree, on the strain rate. No significant influence was found of the initial grain size on the size of the dynamically recrystallized grain after plastic deformation. The existence of an exponential dependence between the average area of recrystallized austenite grain and the deformation temperature, as well as of an involutive dependence on the Zener-Hollomon parameter Z was shown.

5. Acknowledgment

The present work was supported by the Polish Ministry of Science and Higher Education under the research project No 7 T08A 038 18.

6. References

Bywater K.A. & Gladman T. (1976). Influence of composition and microstructure on hot workability of austenitic stainless steels, *Metals Technology*, Vol. 3 (1976), pp. 358-368

Cwajna J.; Maliński M. & Szala J. (1993). The grain size as the structural criterion of the polycrystal quality evaluation, *Materials Engineering*, Vol. XIV (1993), pp. 79-88

Ducki K.J.; Hetmańczyk M. & Kuc D. (2006). Quantitative description of the structure and substructure of hot-deformed Fe-Ni austenitic alloy, *Materials Science Forum*, Vol. 513 (2006), pp. 51-60

Ducki K.J. (2010). Microstructural aspects of deformation, precipitation and strengthening processes in austenitic Fe-Ni superalloy. *Monograph. Copyright by Silesian University of Technology* (2010), pp. 1-136, ISBN 978-83-7335-721-1

Hadasik E. (2005). Methodology for determination of the technological plasticity characteristics by hot torsion test. *Archives of Metallurgy and Materials*, Vol. 50 (2005), pp. 729-746

Hansen N. (1998). Microstructure and properties of deformed metals, *Materials Engineering*, Vol. XIX (1998), pp. 108-115

Härkegård G. & Guédou J.Y. (1998). Disc Materials for Advanced Gas Turbines, Proceedings of the 6th Liége Conference: *Materials for Advanced Power Engineering*, 1998, pp. 913-931

Head A.K.; Humble P.; Clarebrough L.M.; Morton A.L. & Forwood C.T. (1973). Computed Electron Micrographs and Defects Identification, In: *Defects in Crystalline Solids*, Amelinckx S., Gevers R., Nihoul G. (Ed.), 1973, pp. 39-47

Klaar H.J.; Schwaab P. & Österle W. (1992). Round Robin Investigations into the Quantitative Measurement of Dislocation Density in the Electron Microscope, *Praktische Metallographie*, Vol. 29 (1992) (1), pp. 3-26

Kohno M.; Yamada T.; Suzuki A. & Ohta S. (1981). Heavy disk of heat resistant alloy for gas turbine, *Internationale Schmiedetagung 1981, Verein Deutscher Eisenhüttenleute*, Düsseldorf, Vol. 12 (1981), pp. 4.1.1-4.1.22

Koul A.K.; Immarigeon J.P. & Wallace W. (1994). Microstructural control in Ni-base superalloys, In: *Advances in high temperature structural materials and protective coatings*, National Research Council of Canada, Ottawa, 1994, pp. 95-125

McQueen H.J. & Ryan N.D. (2002). Constitutive analysis in hot working, *Materials Science and Engineering*, Vol. A322 (2002), pp. 43-63

Schindler I. & Bořuta J. (1998). Utilization Potentialities of the Torsion Plastometer. *Published by Department of Mechanics and Metal Forming, Silesian University of Technology* (1998) pp. 1-106, ISBN 83-910722-0-7

Sellars C.M. (1998). Role of computer modelling in thermomechanical processing, *Materials Engineering*, Vol. XIX (1998), pp. 100-107

Szala J. (1997). Computer program Quantitative Metallography, *Edited by Department of Materials Science, Silesian University of Technology* (1997)

Zener C. & Hollomon J.H. (1944). Plastic flow and rupture of metals, *Transactions of the ASM*, Vol. 33 (1944), pp. 163-235

Zhou L.X. & Baker T.N. (1994). Effects of strain rate and temperature on deformation behaviour of IN 718 during high temperature deformation, *Materials Science and Engineering*, Vol. A177 (1994), pp. 1-9

7

Recrystallization: A Stage of Rock Formation and Development

R.L. Brodskaya and Yu B. Marin

Saint-Petersburg State Mining University, Saint- Petersburg
Russia

1. Introduction

The goal of the paper is to show the place and mechanism of recrystallization in the complicated and long-term rock evolution. Theoretical preamble to the study, research methods and their results are discussed.

2. Theoretical recrystallization process model

The rock, like any complex system undergoes several significant stages during its development. To each stage of development in time corresponds its own physiographic expression. Let us recall that rock physiography depends on its texture and structure, i.e., relative amount of minerals in the rock, relative and absolute size of mineral grains, their mutual arrangement, orientation and distribution in space. All of these characteristics describe the structure of mineral aggregate (including rock).

We call development stages of mineral individua and aggregates as stages of their ontogenetic development by analogy with the evolution of biological organisms: initiation, growth, and destruction. Inherently, the transition from one stage to another cannot be gradual or smooth. There must be an interval fixed in time between these stages. There are large taxa of rock evolution: effusive, vein, intrusive, orthometamorphic... Rate and duration of crystallization and the formation of magmatic and metamorphic bodies are key evolution factors in this series. Continuing parallels with biological evolution, it is possible to assume that this series corresponds to rocks phylogenesis within one family. Then, for example, basalt - dolerite - gabbro constitute one series of basic rock evolution (phylogenesis). In gabbro mass, there are always mineral aggregate areas that correspond to the processes of late- or post-magmatic alteration, which result in the emergence of new textural and structural relationships in the mineral aggregate. One can observe different development stages of one mineral aggregate. This is our understanding of the difference between ontogenesis and phylogenesis as applied to the rocks.

Rock alterations during its evolution can be recorded at different levels of organization - isotopic, geochemical, mineral. Changes in a mineral aggregate or a real rock that correspond to a certain stage of its development, correspond to the stage of its ontogenesis. Mineral level of investigation taken by the authors assumes that the rock can be polymineral or monomineral, but it is always polycrystalline natural formation, natural mineral aggregate (volcanic glass in this case is not considered). Mineral crystals are formed and exist under conditions of an assembly, collective growth and functioning.

In magmatic rocks, owing to specific character of the crystallization substrate, its dynamic properties and the volume of the crystallization, kinetic characteristics of crystal formation are inconsistent in time. Mineral crystallization takes place under different conditions. This affects the morphology of resulting crystals, their intergrowths and spatial distribution. Usually, mineral crystals in the rock are called "grains". In metamorphic rocks, all transformations proceed in solid state. Dynamic geological conditions associated with new processes cause changes in the structure and composition of mineral aggregates. In magmatic rock, the rate of mineralization reactions and, consequently, crystal growth usually decreases from first portions of the crystallization to last ones. In metamorphic rocks, kinetic inversions in process parameters are possible both towards the increase in the mineral formation rate and towards the decrease. It is reflected in the increase in the mineral grain intergrowth boundaries area, i.e., the roughness of the boundaries increases. Then the process can follow different scenarios. One of them is the granulation of mineral grains, decrease in their size. Another way of system development is the formation of new grains in the area of inequilibrium boundaries intergrowth. This phenomenon is known to material scientists as "mechanical hardening". The formation of new generation individua that absorb "excess" local energy. With another set of circumstances, the totality of energy loading may be beyond the elastic and plastic deformation of the crystal assembly that can result in brittle deformation of solid bodies. One of the thermodynamic process scenarios after the selection of the way of development by the system is its "straightening" in the course of time. The mineral aggregate is adapted to this choice of the system by the flattening of its internal boundaries, i.e., migration of individual sections of the boundaries to a plane parallel to the plane crystal structure grid, which energy corresponds to local potential of the mineral aggregate system in the intergrowth boundary area at the given stage of its development.

Rock physiography in the accepted hierarchy of consideration is a mineral sublevel created by morphology of mineral grains or boundaries of their intergrowth. Mineral individua or grains exist within internal boundaries of a mineral aggregate and differ from one other in different internal structure: some grains are zonal, others contain mineral and/or fluid inclusions, low-angle misorientation of individual blocks of the crystal lattice, etc. Using biological terms, it is possible to say that grains of one mineral in the mineral aggregate can be of different anatomy. It is clear that different anatomy of mineral individua is due to different conditions of their formation, including growth and dissolution in different kinetic regimes.

Changes in the texture and structure of mineral aggregate, as well as the coexistence of mineral grains with different internal structure, are closely related to changing geological conditions of their formation and existence. The "geological conditions" are some external (with respect to the aggregate) physical fields, their energy, forces and orientation such as areas of tectonic stress but occurring within elastic deformation of minerals and rocks, the area of heating from fluid flows located and crystallized near magmatic bodies, etc. If external fields of force are changed, the internal energy of the mineral aggregate must come into compliance with the external energy. In the balancing process, the structure of the mineral aggregate adapts the whole system of mineral grains (mineral aggregate) to new conditions. The adaptation of the assembly of grains is due to the adaptation of the framework of their boundaries, by changing the composition and energy of the boundaries, i.e. by changing the orientation and the area of mineral grain intergrowth boundaries in the aggregate, by changing intergrowth matrix. It is necessary to remind here that the internal energy of the grain assembly consists of the energy of crystal lattices of mineral individua and the energy of mineral grain intergrowth boundaries. Also, we would like to remind that in the massive mineral aggregate, individuum intergrowth boundaries are boundaries of the individua, i.e., as a whole, they comprise the morphology of each mineral individuum and the framework of internal boundaries of mineral individuum aggregate.

One of initial processes of the framework adaptation of aggregate internal boundaries to the changed conditions is its recrystallization, sometimes accumulative recrystallization. In Russian geological literature, it is common practice to call the process of changing the size and shape of mineral grains in the solid state the recrystallization, but there are two types of recrystallization – one with decrease in the size of mineral individua (it is called *recrystallization*) and another with increase in the medium-sized grains (in Russian literature "perecristallizatsia" or *"overcrystallization"*). We would like to repeat that the grain boundaries migration changes the texture of the mineral aggregate. The accumulative recrystallization controls the structure of the mineral aggregate. The mineral individuum boundaries change the orientation relative to the crystal lattice of mineral grain and possibly in space, taking the position that provides them with such an amount of stored energy that can save the grains under new geological, i.e., thermodynamic and kinetic conditions.

However the grain boundaries migration to a stable state under new conditions requires an initial impulse to overcome stable nonequilibrium. A heat flow from approaching or crystallizing intrusion or a fluid flow either an energy flow of tectonic nature can serve as such an impulse. Not only a new compression or stress can be such an impulse, but the decompression as well. In this case, the system of mineral individua adapts to new growth conditions of individuum well-oriented in a new field of force, or the process of accumulative recrystallization. The authors believe that good orientation of the mineral individuum in the field of force is when an individuum occupies a position when the most stable, i.e., the most atomically dense mineral individuum face occurs normally towards the acting stress. Probably, the schistose structure of mica schist forms in such a manner. It is quite possible that this phenomenon is the cause of gneissose structure. It is not inconceivable that that the interaction of external fields of force and aggregate mineral grains hinders the realization of the described scenario. Then the aggregate adaptation will involve the accumulative recrystallization process. Grains of one mineral form glomero-grained clusters, i.e., subaggregates consisting of grains of one mineral. But there were cases when generated monomineral subaggregates formed rather stable distinct boundary between the subaggregate and grain matrix in the aggregate. The monomineral subaggregate attains crystal-like morphology, i.e., a shape when part of its boundaries with mineral aggregate look like simple forms inherent in this mineral. The process of levelling, balancing of intergrowth energy of mineral individua (recrystallization) continues inside the subaggregate. (Fig. 1, 2, 3)

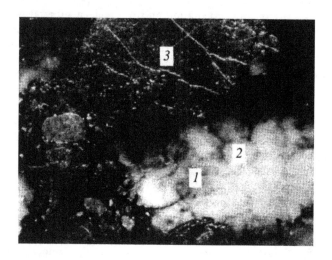

Fig. 1. Fragment of conglomerate from Carbon–Lider ridge. 1- quartz pebble; formation stage of inner boundaries of subaggregate, which is relevant to grain faces - attractor(2); 3 – pebble of pyrite grain. 1.5x.

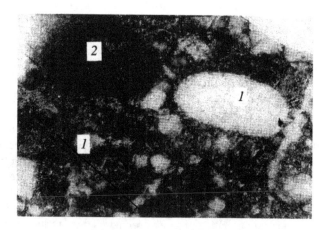

Fig. 2. Fragment of conglomerate from Ventersdop ridge. 1 – pebble of quartz of 1 kind; 2 – quartz pebble of 2 kind (amoebic contours, in the center quartz are free from inclusions, chlorite micrograins paint margin to dark-green color). White points on the right – accumulation of fine grained pyrite. Gray angular segregations are phyllite. 1.2x.

Fig. 3. Fragment of conglomerate from Carbon–Lider ridge. Cataclastic quartz pebble of 1 kind; 2 – pebble of third kind, composition and color is similar to basic matrix. White mass around pebble – aggregate of pyrite. 1,5x.

External flow or the initial impulse induces the energy flow from each mineral grain. This is the energy of edge dislocation of mineral individuum, energy of its boundaries. Energy of dislocations and defects in the crystal lattice of each mineral individuum is involved in the general flow. Thus, the stable equilibrium becomes unstable. Trace elements located in defects and dislocations migrate from their places together with induced energy flow. The flow is directed towards the edge dislocation of mineral grain – its boundary. This energy and its flows provide the grain distillation from trace elements, mineral and fluid inclusions, subboundaries – low-angle boundaries within the crystal lattice (e.g., subboundaries between blocks of cloud extinction in quartz, "loops" and "oblique walls" in olivine). When the impulse energy is sufficient, the process of solid solution disintegration is being formed. Many minerals represent such solid solutions of one mineral in another one. The process of solid solution decomposition results in the appearance of specific, easily recognizable decomposition structure (Fig. 4 (a, b). This is a new instability, which activates migration of subboundaries within the grain. It's possible to indentify by means of displacement character of subboundaries and stimulated movement forces two types of recrystallization - rotational and migrational (Fig. 5,6). This is the way of changing the anatomy of mineral individua; this is the way of replenishing the impulsive force energy for the formation of new mineral grain boundaries. Accumulative overcrystallization within the grain, i.e., the aggregation of micro- and nano-individua clusters takes place simultaneously with the migration of boundaries. Trace elements and newly formed mineral phases are "squeezed-out" to grain boundaries to generate their own mineral form with its own boundaries. This is the mechanism of overcrystallization and mineral formation at grain boundaries in the solid state (Fig.7).

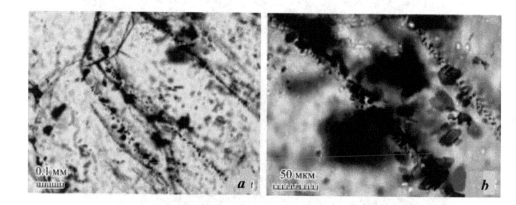

Fig. 4. (a,b) Structures of the solid solutions decomposition in olivine; the newly formed phase is chrome spinelide. Dunite, Gulinsky massif; photographs in transparent light, without analyzer

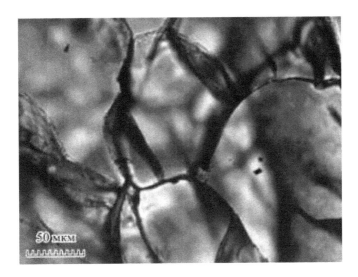

Fig. 5. Rotational recrystallization of the olivine aggregate from dunite of Gulinsky massif; photograph in transparent light, with analyzer.

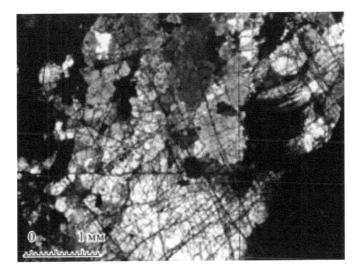

Fig. 6. Migrational recrystallization of the olivine aggregate from dunite of Galmoenansky massif; photograph in transparent light, with analyzer.

The surface energy of mineral grain edge dislocations and the boundaries of their intergrowth in the aggregate is, as already mentioned, an instrument in the mechanism of balancing between the internal energy of the aggregate and the external energy of the field of force under changing geological conditions. The amount of the mineral grain surface energy consists of the edge dislocation energy of the crystal lattice of mineral individuum and the presence of some admixtures, i.e., first of all, depends on the orientation of the boundary (edge dislocation) relative to the individuum crystal lattice (Fig. 8 (a, b, c). However, main role in boundary migration is traditionally given to the specific energy of the surface area rather than to the surface energy.

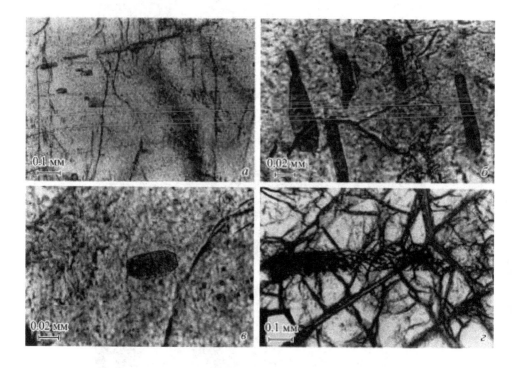

Fig. 7. Forming of new minerals in deformational substructures of olivine (Arai Shoji, 1978). Photographed with different magnifications in transparent light

Fig. 8. (a, b, c). Regularly oriented lamellae of chrome spinelide and the skeleton inclusions of spinel in the olivine grain. Dunite, Gulinsky massif; photographs in transparent light, without analyzer, in different thin sections

Increase in some components of the external field of force (stress, lithostatic pressure, fluid flow pressure, heat flow) results in the increase in the amount and energy of internal boundaries of mineral aggregate. This process inspires the increase in the boundary density in the mineral aggregate space as well as the specific surface energy. The decrease in the external energy flow necessitates the decrease in the internal energy of the mineral aggregate. To this change in external geological conditions the aggregate is adapted due to the decrease in the surface energy of mineral individua. In the first case (increase in the field of force) porphyraceous structures form, in the second - monomineral subaggregates. The selection of the development path system depends on the necessity to decrease the internal energy. This is possible owing to increase in the area of mineral individuum boundaries and decrease in its specific surface energy. This is also possible due to the decrease in the grain intergrowth energy of one mineral. In polymineral aggregate, the least amount of the energy is absorbed by intergrowth boundaries of one mineral.

At grain boundaries of one mineral, the intergrowth energy is lower than in intergrowths of different minerals. Most likely, just the "energy benefit" is the motivation of the accumulative recrystallization that covers vast mineral aggregate spaces. This is the way of formation of glomo-grained subaggregates within the massif aggregate matrix with regular grain distribution of all the minerals. Which mineral in the "struggle for survival" will decrease or increase the density of its boundaries depends on marginal conditions of existence of this or that mineral in the aggregate. It is quite possible that material supply into the crystallization system can result in the generation of such conditions (Fig. 9).

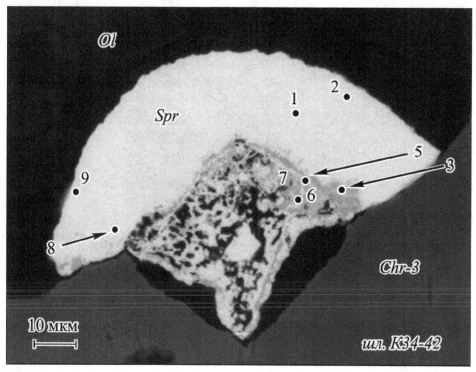

Fig. 9. Subaggregate of grains of platinum and iridium arsenides at the boundary of intergrowth with subaggregates of the third generation chromite (*Chr-3*) and olivine; numbers of points correspond to microprobe analyses: *1, 2, 8, 9* — sperrylite, *5, 6* — raresite (transparent polished section K-34-42); electron microscope.

The character of subaggregate boundaries is an important kinetic feature of the occurring processes. It may be flat or irregular to variable degrees. Under some conditions, the migration of grain boundaries results in their straightening and integration into intergrowths, i.e., increase in the grain volume.

Studying the energy models of internal boundaries relationship in the mineral aggregate shows that changed position of mineral individuum intergrowth boundaries or recrystallization is a response of the mineral grain assembly to new geological conditions under which the mineral aggregate occurs. Main tool of aggregate adaptation to new conditions is the migration of individuum intergrowth boundaries to such a position in space and relative to the crystal lattice of each of the intergrown grains that ensures its stability in the field of changing force and energy. It is quite possible that not all boundaries and not all minerals have to migrate when they change their orientation in space. Thus, atomically dense boundaries, i.e., simple-shape faces are the most stable mineral individuum boundaries. Boundaries of mineral grains, which in crystallographic coordinates have sufficiently high symbols, i.e., have relatively low store energy, are capable of migration in the aggregate space. It is natural that the boundary migration is possible until they reach equilibrium state with one another. But other situations are also possible. For example, if recrystallization accompanies metasomatosis with additional supply of some material. In this case, growth of this material from new crystallization centers is possible either the growth due to the increase in mineral individuum volume if its composition corresponds to that of the supplied material. Porphyroblast growth is possible when supplied material flow rates are rather high.

3. Mineral aggregate ontogenesis and physiography

Earlier it was said that the necessity to decrease the internal energy of mineral aggregate during its adaptation to the external energy (e.g., while decompression) results in the change of the structure of the mineral grain assembly – accumulative recrystallization. Probably the spotty structure of phyllite slate is a result of this process. Quartz grains are accumulated into such glomero-grained subaggregates characterized by irregular distribution and rather variable volume in the rock space. High rate of changes in external conditions can result at some stage or other in the arrangement of a boundary between a monomineral glomero-grained unit and aggregate matrix. In this case, the external boundary is formed by the combination of identical boundaries of mineral grains located at edges of glomero-grained bundles. Individuum boundaries usually correspond to a simply-shaped face of this mineral with relatively high stability factor under given kinetic conditions. The authors came to this conclusion while studying texture of the so-called quartz pebbles in auriferous reefs of Witwatersrand. Macroscopic investigations of samples from conglomerate outcrops of the Carbon Leader and Ventersdorp reefs, suggest the similarity of the morphology of "pebbles" and that of quartz crystals due to the presence of surfaces in the pebble that resemble prism face. Thin sections were made of several quartz "pebbles", which had in their faceting external prism "faces". In the quartz grain subaggregate, microscopic investigations revealed the presence of sections, which boundaries were similar to those of external morphology of the pebbles. The authors interpreted this fact as an existence of a grain-attractor within quartz individuum subaggregate. It is just these boundaries are the most stable for quartz, and therefore most beneficial for the conservation of quartz subaggregate in the regime of unstable parameters of changes in geological conditions, which also imply thermodynamic ones. Most likely, this is the alternation of compression and decompression in the course of compression and extension of host rocks during hydration and dehydration of intergranular space of mineral aggregate. Occurrence of hydrofilms in the mineral aggregate not only increases the plasticity of the rock as a whole, but also helps to increase the resistance of mineral individua to elastic and plastic deformations. Under certain conditions, such quartz monomineral subaggregates can be transformed into blastoporphyric quartz "crystals" if quartz individua will be able to adapt to each other not only by boundaries of appropriate density, but also due to coherent orientation of crystal lattices of porphyroblasts. Such examples are recorded not only in Witwatersrand, but also among porphyraceous dunite aggregates of the Inagli, Galmoenan massifs. It is interesting that in these cases, grains-attractors with orientation and morphology of the subaggregate are also observed inside olivine subaggregates.

4. General procedure of quantitative ontogenetic analysis

Stereometric analysis of rocks and ores made in thin sections and polished sections was the main method of implementing the above mentioned ideas. Quantitative assessment of the parameters of the structure allows the estimation of the recrystallization degree of the aggregate and mineral individua. Structural characteristics become parameters of the texture and structure because of the application of crystallographic and topological methods of analysis. Anatomy, i.e., internal structure of mineral individua also has its quantitative measure. Density of subboundaries, density of fluid and mineral inclusions and other features of refining mineral individua during recrystallization are estimated here.

This research trend (quantitative ontogenetic analysis) allows unbiased assessment of the stage of mineral aggregate evolution by distinguishing individual generations and paragenetic (simultaneous) associations of mineral generations. Not only is the history of mineral aggregate restored, but a place of mineralization in ontogenesis as well.

Thus, the main method of mineral aggregate structure interpretation is the ontogenetic analysis of the mineral aggregate and its individua. Main method of implementation of the expressed ideas is stereometric analysis of rocks and ores. A procedure of studies using polished sections and thin sections was elaborated. A representative area is necessary to obtain metric assessment of the mineral aggregate structure under the microscope. Standard area of thin sections is used, but the design of the integration device MIU-5M allows analysis of thin sections with an area of no more than 40×40 mm. Analysis sensitivity is not worse then 4 μm. It means that using scanning table, it is possible to get grid coordinates of points occurring within the specimen plane spaced at a distance of 8-10 μm from one another. Quantitative measure of fabric parameters allows evaluation of degree of recrystallization of the aggregate and mineral individua. Structural characteristics become texture/structure parameters due to simultaneous employment of well-known crystal-optic, crystallographic, geometric, and topologic analytical methods. Anatomy of mineral individua has quantitative measure. Following features are estimated here: subboundary density within mineral grain, density of fluid and minerals inclusions and other features both residual, primary, and refinement features of mineral individuum anatomy while recrystallization. Roughness of intergrowth boundaries of all mineral grains or grains of one mineral in the aggregate either grains of individual mineral generations can be changed and calculated using several methods.

Such parameters of mineral aggregate fabric as total area of internal boundaries of mineral aggregate, modal portion of individuum boundaries of each mineral, modal and normative granulometric compositions, character of mineral grain distribution in aggregate or frequency index of individuum intergrowth of one mineral with grains of other minerals, etc. are measured simultaneously while scanning a specimen (thin section or polished section). These are the so called integral characteristics of rock structure. Frequency characteristics of mineral grain boundaries can be obtained in the course of their analysis using the device of fractal dimensions or Fourier harmonic decomposition. All quantitative characteristics and fabric parameters necessary for ontogenetic analysis of mineral aggregates are real functional capabilities of the Mineralogical Integration Device (MIU-5M in Russian). The elaborated procedure enables to get and use 22 fabric parameters or part of them in any combination and amount.

This trend of studying rocks and ores can be named quantitative ontogenetic analysis. Its use in investigating thin sections and polished sections allows unbiased assessment of stage of mineral aggregate evolution and identify separate generations and paragenetic (simultaneous) associations of mineral generations. Not only the history of the mineral aggregate, but also a place of mineralization in the ontogeny is reconstructed in the transformations sequence of mineral individua assembly.

5. Conclusions

We discussed the recrystallization process in a mineral aggregate as a migration process of mineral individuum boundaries, mineral grain intergrowth boundaries, as the process of changes in the framework of internal boundaries of the aggregate. Migration reasons were formulated as a mechanism of mineral aggregate adaptation to changed (as compared to initial conditions of formation) geological conditions of rock existence. On the way of aggregate adaptation to changing geological conditions, the recrystallization, similar to accumulative recrystallization, is possible at all levels of mineral matter existence.

6. References

Arai Shoji. Chromian spinel lamellae in olivine from the Iwanai-dake peridotite mass, Hokkaido, Japan /Earth and Planetary Science Letters. 1978. N 39. C. 267 – 273.

Brodskaya R.L., Shumskaya N.I. // Transactions of the USSR Ac. Sci.. 1998. V. 362. No. 3, pp. 378-381(in Russian)

Brodskaya R.L., Bilskaya I.V., Kobzeva Yu.V., Lyachnitskaya V.D., Rachmanova N.V., Talovina I.V. Formation of surface and properties of the mineral individual borders in aggregate destruction. IGC, 2000. Brazil.sec. of phys. and chem. minerals, publishing in CD.

Brodskaya R.L., Marin Yu.B. Problem of Internal Structure Modeling of Ordered and Equilibrium Mineralogical-Petrographic Systems// ZVMO 2001. P. CXXX. No. 6, pp. 1-14 (in Russian)

Brodskaya R.L., Bilskaya I.V., Kobzeva Yu.V., Lyachnitskaya V.D. Typomorphic Structural Features of Ultramafite Mineral Aggregates and Mechanism of Chrome Spinellide Concentrations in them // ZVMO. 2003.P. CXXXII. No. 4, pp. 18-37 (in Russian)

Brodskaya R.L., Marin Yu.B. Rock Structuring: Adaptation Mechanism of the System to Inequilibrium Thermodynamic Processes /Collection of Articles "Rocks". 2004. Apatity, pp. 19-26(in Russian)

Brodskaya R.L., Bilskaya I.V., Lyachnitskaya V.D., Markovsky B.A., Sidorov E.G. Formation of PGE Mineralization in Ultramafite of the Galmoenan Massif (Koryakia) // Transactions of the 8th International Conference "New Ideas in Geoscience". Moscow. 2007. V. 5, pp. 37-39 (in Russian)

Brodskaya R.L., Marin Yu.B. Formation Model of Mineral Aggregate Internal Boundaries and Examples of Its Application. Transactions of the 8th International Conference "New Ideas in Geoscience". Moscow. 2007. V. 3, pp. 56 – 59 (in Russian)

R.L. Brodskaya, I.V. Bil'skaya, V.D. Lyakhnitskaya, B.A. Markovsky, E.G. Sidorov. Boundaries of Intergrowths between Mineral Individuals: A Zone of Secondary Mineral Formation in Aggregates.// Geology of Ore Deposits. 2007. Vol. 49. No. 8, pp. 669-680. © Pleiades Publishing, Ltd.

R.L. Brodskaya, I.V. Bilskaya, B.A. Markovsky. Ontogenic Analysis of Individual Olivine
    Grains in Ultramafic Rocks. // Geology of Ore Deposits.2010. Vol.52. No.7, pp.
    566-573.

Ponomarev V.S. Energy Saturation of Geological Environment. Transactions of Geological
    Institute. 2008. Moscow. Nauka. Issue 582. 379 p.(in Russian)

Shcheglov A.D., Shumskaya N.I. // Transactions of the USSR Ac. Sci. 1995. V. 340. No. 5,
    pp. 667-671(in Russian)

Witwatersrand Gold – 100 Years/ Ed. by E.S. Androbus. Geol.Soc.S.Africa, 1986.P. 298.

Part 2

Recrystallization in Pharmacology

8

Recrystallization of Drugs: Significance on Pharmaceutical Processing

Yousef Javadzadeh, Sanaz Hamedeyazdan and Solmaz Asnaashari
Biotechnology Research Center and Faculty of Pharmacy,
Tabriz University of Medical Sciences
Iran

1. Introduction

Not surprisingly, the wide range of effective medicinal agents available today is one of the greatest scientific achievements. Regardless of the advancements in effectiveness and safety of the medicines embedded in dosage forms, the pharmaceutical concept of the latter is growing to be ever more eminent (Adibkia et al., 2011). Following on from recent advancements, in a time of increased considerations to the level of sophistication in designing pharmaceutical dosage forms keeping pace with advances in drug discovery methods, it seems as important as ever to study the physicochemical properties of active pharmaceutical ingredients, prerequisite for a successful product formulation.

As far as we know, the molecular structure of any drug compound typically defines all of its physical, chemical and biological actions. Owing to the fact that a certain kind of drug might be offered in a variety of solid forms, including polymorphs, solvates, hydrates, salts, co-crystals and amorphous solids, the choice and design of the ideal solid-state chemistry of the pharmaceutical solid form would be critically important to a superior drug development. Accordingly, drug crystals could be modified in different ways including recrystallization, which would affect the physical and physicochemical properties such as melting point, solubility, true density, drug release profile, flowability and tabletability of the pharmaceutical dosage forms (Harbury, 1947; JamaliMitchell, 1973; Jozwiakowski et al., 1996).

Recrystallization is a simple and inexpensive method for scaling up the drug developments to a commercial level. Significant advances in the different pharmaceutical dosage form technologies renders drug recrystallization as a green technique due to the savings of costs, time, energy and less machinery as well as fewer personnel. Recrystallization is one aspect of precipitation obtained through a variation of the solubility conditions and the amount of dissolved solute in an increased temperature. In general, production of another crystalline from of a drug and also purification are the two major sets of applications for drug recrystallization processes. Briefly, in a drug recrystallization process, a hot saturated solution of the drug is prepared with only enough solvent to dissolve it at the boiling point of the solution. Once the solution is cooled the purified drug component or a new crystal form of the drug separates as a result of the lower solubility of that crystalline form of a drug in the respective solvent at lower temperatures.

Since impurities are present in fairly small amounts of drug solutions they do not crystallize in recrystallization and they are ready to separate from the formed drug crystal (Tiwary, 2001). For instance, in the case of natural medicinal compounds obtained from natural sources which almost always contain impurities, in order to obtain a pure drug, usual major steps in the recrystallization process have been schematically demonstrated in figure 1. Purifying a sample drug of compound X which is contaminated by a small amount of compound Y, would be established with an appropriate hot solvent in which all of compound Y is soluble at room temperature and the impurities will stay insoluble in and pass through filter paper, leaving only pure drug crystals behind, as has been shown in figure 1.

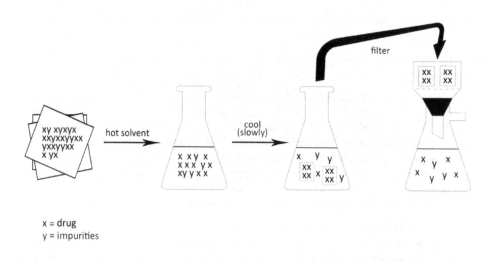

x = drug
y = impurities

Fig. 1. Schematic protocol of drug recrystallization for further purification.

It is worth mentioning that in any recrystallization technique some drug loss is inevitable and the total recovery would be less than 100%, seeing that even at the lower temperatures the target drug has some finite solubility in the solvent and is lost subsequently when solvent and soluble impurities are removed. Moreover, selection of the right solvent in the recrystallization seems to be one of the crucial features of the process and is made on a case-by-case basis (MirmehrabiRohani, 2005; ChenTrout, 2008). This is because of solubility variation of different drug compounds in different solvents so that a certain drug not only should have the highest solubility in the solvent of choice at its boiling point, but also it should show a markedly diminished solubility at lower temperatures of the same solvent. Although recrystallization is a very common technique used to purify drugs, it has a basic limitation for the compounds that are mostly pure and other techniques of separations are of use for the drug mixtures containing several major components which could not be purified by recrystallization methods.

Ever since, the leading physicochemical properties of a unique form of a drug could seriously influence the bioavailability, manufacturability purification, stability, solubility, and other characteristics, identifying these potential liabilities allows us to predict, control and avoid any complexities that may arise during drug development stages (Krishnaiah, 2010). This would be beneficial in preventing the drug development efforts to costly late stage product failures throughout the manufacture and storage periods too. However, in order to fully control the crystallization process, the link between a particular solid form of a drug structure and its functional physicochemical properties, still challenges to be better established to facilitate the suitable drug production. A large and growing body of literatures has been published on recrystallization techniques and physicochemical properties of a drug. Herein, we gathered some of the related reports of drug recrystallization to have an overlook on the crystal habit of a drug on some basic physical properties of pharmaceutical dosage forms signifying how these factors are interrelated.

2. Impact of crystal habit on pharmaceutical processing

Drug discovery and characterization relies on the nature of the target molecule and the relative physicochemical properties of drugs. Identifying all relevant crystal habit of a drug which is an important variable in pharmaceutical manufacturing at the development phase from research to commercialization is of substantial value. Due to the different crystal form variations of some basic physical properties like, solubility, dissolution rate, melting behavior, and certain micromeritic properties or performance characteristics, e.g. tablet compressibility, mechanical strength, powder flow provide alternatives to select a form that presents the suitable balance of critical properties for development into the drug product. Establishing such modification information at an early stage of drug development process lessens the risk of process alterations given form changes and brings in the opportunity to attain more comprehensive rational property coverage.

The merit of changes in crystal surface form and habit of drug powders by recrystallization method is much more realized when there is an essential to diminish variations in raw material characteristics, to certify reproducibility of results during drug preformulation, and also to judge fairly about the cause of poor performance of a dosage form. Besides, the changes in crystal habit of a drug going together with or without polymorphic transformation at some point in processing storage could account for serious implications of physical stability in dosage forms. Thus, it seems underlying to have a deeper insight to the crystal structures and control the solid-state chemistry of drug substances to design a more systematic and intellectual pharmaceutical dosage forms.

In a survey carried out by Sinclair et al. ibipinabant a potent and highly selective cannabinoid receptor antagonist was evaluated for its solid-state physical stability and recrystallization kinetics in tablet dosage forms using fourier transform raman spectroscopy. The findings of the study showed that exposure to moisture had notable influence on the crystallinity of amorphous ibipinabant. The recrystallization kinetics measurements revealed a two-step process with an induction period (nucleation) followed by rod-like crystal growth by application of the Johnson–Mehl–Avrami kinetic model. On the whole their method provided reliable and highly accurate predictive crystallinity assessments after exposure to a variety of stability storage conditions for ibipinabant (Sinclair et al., 2011).

Recently, Dahlberg et al. analyzed the stability of the amorphous drug, flutamide, by a combination of localized nuclear magnetic resonance (NMR) spectroscopic and NMR imaging techniques. Owing to the fact that, NMR relaxation is sensitive to both the crystalline and amorphous state and the size of the drug substance, it allows for an in situ monitoring of the state of the drug during tablet disintegration and dissolution periods. With regard to the results of the NMR experiments, recrystallization was believed to be related to its enabling factors such as local hydration level and local mobility of the polymer matrix. Eventually, it was verified that the primarily amorphous flutamide may recrystallize either by nanoparticle coalescence or by ripening of crystalline particles (Dahlberg et al., 2011).

The solid-state properties of sulfathiazole and chlorpropamide were modified through recrystallization using supercritical antisolvent process by Yeo et al. They confirmed that the operating conditions of the system such as carbon dioxide injection rate, type of solvent, and temperature significantly had an effect on the physical characteristics of the resulting crystals. Considering the results of the study, drug crystals processed with supercritical system exhibited more ordered appearances with clean surfaces and sharp angles compared with the unprocessed particles where crystal habit changed from tabular to acicular when the carbon dioxide injection rate increased. Photomicrographs of sulfathiazole crystals with methanol as a solvent, confirmed a needle-like acicular and a tabular crystal habits in rapid and slow injections, respectively. Whereas, in the case of chlorpropamide, processed drug particles in the rapid injection experiment exhibited columnar habit in a regular shape, while relatively large crystals with sharp angles were observed in the slow injection mode when acetone was used as the solvent. Overall experimental observations suggested that the supercritical antisolvent process could provide favorable environment for the solid growth of a single type of crystalline drug, minimizing the conditions for growth-related imperfections (Yeo et al., 2003).

According to the fact that thermal analysis has been frequently used to identify crystal forms of drugs and in the course of thermal analysis, crystal transformation is often observed as well as melting and decomposition, Suzuki et al. studied mechanisms of thermal crystal transformation through melting and recrystallization. They characterized two anhydrates (α-from and β-from) and two hydrates (hemihydrate and monohydrate) forms of a novel fluoroquinolone antibiotic, sitafloxacin, in addition to sesquihydrate which is used in the marketed drug products. The results of crystal structural that were characterized by infrared spectroscopy, X-ray powder diffractometry and thermal analysis revealed quinolone rings of sitafloxacin had distorted planar structure and quinolone ring of the drug in α-form and monohydrate hold opposite torsion to those in β-form and sesquihydrate. These kinds of thermal analysis are often recommended as a routine tool for quality control of thermal dehydration and subsequent crystal conversion of drugs (Suzuki et al., 2010).

2.1 Compaction and flowability

Acquiring a clear notion of why certain drug materials are prone to problems during compaction and dominate the relative constraints to offer a successful compaction and tableting strategy of pharmaceutical powders would involve an understanding of the fundamental properties of drug powders. Therefore, it is important to determine the effect of different physicochemical properties such as particle size, shape, surface area, polymorphic form, crystal habit, hydrates, and processing conditions on the compaction of powders. As different crystal habits of a certain drug hold dissimilar planes, they have differing points in their specific surface and free surface energies. Even so, alternative recrystallization solvents could develop crystal with defined crystal habit, size and shape as well as compressibility properties. The nature and amount of these changes count on the recrystallization conditions including the presence of impurities, type of solvents and cooling rates. This view is supported by a variety of papers at molecular level developing the knowledge of solid-state properties such as crystal structure, crystal habit, and polymorphism influence on the mechanical properties of powders in an attempt to identify and modify physical properties of bulk solids of drugs (Liebenberg et al., 1999; Maghsoodi et al., 2007).

Seton et al. evaluated the particle morphology of ibuprofen, an anti-inflammatory drug, by recrystallization from a range of solvents and investigated the following influence on compaction properties. The compaction data achieved from properties of the ibuprofen control and recrystallized samples at different compaction forces and speeds revealed equal or better tablet strength than the control, whilst ibuprofen recrystallized from 2-ethoxyethyl acetate exhibited lower levels of elastic energy during compaction. In addition, the recrystallized ibuprofen samples demonstrated flowablity equivalent quality to the ibuprofen control, excluding the ibuprofen recrystallized from acetone which showed excellent flow properties. Generally, the results displayed ibuprofen recrystallization from various solvents could offer advantages in terms of particle morphology, flowability and compaction properties (Seton et al., 2010).

In another study an anti-epileptic drug, phenytoin, crystals in the form of free acid, having distinct types of habits, was modified via different recrystallization conditions and techniques by Nokhodchi et al. Several sets of experimental conditions for temperature, solvent evaporation and watering-out techniques were applied for evaluation of the drug recrystallization in ethanol and acetone solvents. The solid state characteristics and compaction properties of the crystal habit with factors affecting the resultant crystals were also evaluated. The physical characteristics of the crystals were investigated using scanning electron microscopy, X-ray powder diffractometry, FT-IR spectrometry and differential scanning calorimetry. They confirmed that using watering-out technique as a crystallization method, produced thin plate crystals, while the crystals obtained by other methods were needle shape for alcoholic solutions and rhombic for acetone solutions. Although the crystallization medium had central effect on phenytoin crystal habit modification, altering crystallization temperature had no effect on crystal habits except a change in size of crystals. In the case of compaction, the crystals produced from alcohol or acetone showed high crushing strengths as a result of lower porosity and lower elastic recoveries (Nokhodchi et al., 2003).

As we know, ascorbic acid crystals are unsuitable for direct tableting due to their poorly compactible properties, Kawashima et al. designed spherically agglomerated crystals of ascorbic acid with improved compactibility for direct tableting. They precipitated ascorbic acid crystals by a solvent change method, followed by their agglomerations with the emulsion solvent diffusion or spherical agglomeration mechanism, depending on the solvent combination for crystallization. Considering the results of the study, under static compression, effectively the proper compact with a sufficient strength was produced. After all improved micromeritic properties, such as flowability and packability for the spherically agglomerated crystals were obtained for crystals of ascorbic acid with the spherical crystallization technique (Kawashima, 2003).

Designing a suitable dosage form with an ideal physicochemical and mechanical property is an important basic principle of drug delivery systems. As follows, crystal structure, shape, and size of drug substances have a huge economical and practical effect at all stages of development from research to commercialization. So, there is a necessity to control the critical properties of drugs for their readiness and capacity to form a tablet which are dominating dosage forms in pharmaceutical dosage form manufacturing.

2.2 Solubility, dissolution, and bioavailability of drugs

Nowadays, in pharmaceutical companies drugs with restricted aqueous solubility have become ever more prevalent and challengeable in the research and development stages. Slow drug dissolution in biological fluids, insufficient and inconsistent systemic exposure and subsequent inadequate efficacy in patients, are some of routine challenges to be coped with during the development of poorly water-soluble drug substances especially when they are administrated orally. Notable numbers of drugs especially new drug candidates are in a biopharmaceutical classification of low solubility (BCS Class II and IV) keeping drug dissolution rate as the limiting factor for the drug absorption and attaining suitable blood-levels of the drugs (LobenbergAmidon, 2000). These inadequacies in solubility of clinically established drug substances in water and in the gastric fluids make problems in drug dissolution rate and oral bioavailability of drugs, as well (Blagden et al., 2007). Consequently, there is a basic requisite to deliver such drugs in a way that gives a chance of sufficient dissolution rate, absorption,and demonstrating suitable clinical efficacy.

Numerous scientific and technological advancements have been made in the research and development for improving and maximizing dissolution rates of the mentioned types of drugs. Despite enhancements in solubility and dissolution rate and oral bioavailability of poorly water-soluble drugs with the customary pharmaceutical technologists, still there are concerns about the success of those methods in the complexities arise from the specific physicochemical nature of the drug molecule itself (Krishnaiah S.R., 2010). One of the thriving trends in enhancing the solubility, dissolution rate and subsequent bioavailability of poorly soluble drugs is to deal with crystal forms of materials which could potentially be applicable to a broad range of drugs with different crystalline habits (Yeo, Kim et al., 2003). On account of many factors such as crystal habit, size and even polymorphic forms of a drug, dissolution rates would enhance through habit recrystallization. A number of reports in the literature validated the effects of crystal morphology variation on solubility, in vitro dissolution rate, holding potentials for improving drug bioavailability (Kawashima et al., 1986; Carino et al., 2006).

Several studies in this filed have shown that exposure of diverse crystal faces determines the nature of the wettability and consequent enhancements in dissolution rate of the drugs with different crystalline shapes (Heng et al., 2006). In 2000, Kobayashi et al. published a paper in which they presented different dissolution rates for carbamazepine, where the dihydrate form of the drug in simulated fluids (pH 1.2) had notably slower dissolution rates than the anhydrous forms (forms I and III). Although the metastable polymorph (Form III) possessed greatest rates of dissolution at the initial stages, reductions in dissolution rate at later time points of the profile was achieved due to the rapid conversion of metastable polymorph (Form III) to the dihydrate. Nevertheless, in another study carried out by Tian et al. the behavior of carbamazepine and dihydrate compacts during in vitro dissolution tests various factors were evaluated. Considering the results, presence of excipients such as polyethylene glycol (PEG) and hydroxyl propyl methyl cellulose (HPMC) inhibited the conversion of carbamazepine to the hydrated form following decreased rates of drug dissolution (Tian et al., 2007). Application of different drug habits in pharmaceutical dosage forms could vary the dissolution rates, as the use of metastable polymorphs in enhancing drug dissolution rates. They also performed bioavailability tests in dogs to determine the effects of physicochemical properties of drug form I, form III and dihydrate on the plasma level of carbamazepine. Similar to the findings of other dissolution studies for carbamazepine, drug bioavailability that had been measured was lowest for the dehydrate form. The lower drug bioavailability established with metastable form was in consistent with the probable conversion of the drug habit to the dihydrate (Kobayashi et al., 2000).

Intrinsic solubility of three crystals of diclofenac, a nonsteroidal anti-inflammatory drug, was investigated by Llinas et al. The crystal habits were characterized and detected by thermo gravimetric analysis, differential scanning calorimetry, and X-ray diffraction. They recrystallized the anhydrous sodium salt of commercially available diclofenac with ethanol and precipitated as a hydrated drug that provided consistent results for the intrinsic solubility. Regarding the broad range of values which have been reported for aqueous diclofenac solubility in the literature, they claimed their solubility records were at the smaller end of the range (Llinas et al., 2007).

Perlovich et al. analyzed four new crystal structures of the sulfonamides by X-ray diffraction experiments and comparative analysis of molecular conformational states and hydrogen bonds networks by graph set notations in the crystal lattices. They established temperature dependencies of the solubility in water, n-octanol as well as thermodynamic functions of solubility and solvation processes for the compounds. According to Perlovich et al. distinguishing between enthalpy and entropy leads to the insight that the mechanism is different for the different molecules where it may be of importance for further assessment of distribution of drug molecules and provide a better understanding of biopharmaceutical properties of drugs (Perlovich et al., 2008).

Dipyridamole as a critical antiplatelet and peripheral vasodilator drug is known to have properties of water insolubility and poor bioavailability which are the limitations of its effectiveness in clinical usage. Adhiyaman et al. characterized dipyridamole crystals with different types of habits by recrystallization from selected solvents. Physicochemical characteristics of the crystals were assessed via scanning electron microscopy, X-ray powder diffractometry, IR spectrometry and differential scanning calorimetry. The developed crystals of dipyridamole under optimized conditions ensue in different crystalline habits that significantly improved dissolution rate compared to original dipyridamole. Recrystallized dipyridamole with benzene and acetonitrile, produced needle shaped crystals and the ones recrystallizaed with methanol produced rectangular shaped crystals. Whereas smooth needle shaped crystals were obtained with the methanolic solution of the drug in the presence of Tween-80, Povidone K30 and PEG-4000 (AdhiyamanBasu, 2006). Generally, these results were in consistence with the possibility of controlling and enhancing the drug release properties following by a probable improvement in bioavailability of drug particles through characterization of drug crystals.

Recrystallization of phenytoin in ethanol and acetone by Nokhodchi et al. was shown to produce needle-like and rhombic crystal habits which brought about identical dissolution rates of crystals obtained from both solvents. Considering the results, nature of recrystallization solvents in this case had no effect on dissolution profiles. It was suggested that the differences in dissolution rates for phenytoin was related to the surface area of various crystals with different shapes (Nokhodchi, Bolourtchian et al., 2003). Wettability and the changes in intrinsic dissolution rate of doped phenytoin crystals were evaluated by Chow et al. They stated that the differences in dissolution rates of phenytoin crystalline powders with different morphology were mainly because of the changes in surface area rather than the improvements in the wetting of more polar surface moieties. However, they stated that the areas of the relatively polar faces seem to be valuable determinants of the drug release profiles of doped phenytoin crystals along with the correlation of the surface tensions (Chow et al., 1995).

Talinolol is a cardioselective beta blocker agent that is known to have different crystal structures with strongly differing solubilities when pure water, acetate, or phosphate buffers are employed as dissolution media. Wagner et al. have studied the impact of different dissolution media controlling the crystal structures of talinolol influencing the dissolution rate and solubility of the drug. The crystal structures were analyzed by means of light microscopy, differential scanning calorimetry, and X-ray powder diffraction, detecting the variations of talinolol crystal structures being the source of incomplete and unpredictable nature of the drug bioavailability (Wagner et al., 2003).

Carbamazepine, a routinely used drug in the treatment of epilepsy and trigeminal neuralgia, exists in four polymorphic forms and as a hydrate which could modify the physicochemical properties of the drug. In our previously published paper we established enhanced physicomechanical properties of carbamazepine via recrystallization at different pH values. The resultant habits of carbamazepine crystals varied from flaky or thin plate-like to needle shape structures which were ascertained using scanning electron microscopy and X-ray powder diffraction. Considering the results of the in vitro dissolution evaluations of carbamazepine samples, a higher dissolution rate for carbamazepine crystals were obtained from media with pH 11 and 1 compared to the original carbamazepine sample. After all, the carbamazepine particles recrystallized from aqueous solutions with different pH values revealed superior mechanical properties which were generally in consistence with the similar studies of drug recrystallization (Grzesiak et al., 2003; Javadzadeh et al., 2009).

In spite of the absolute potential of drug habit modifications in dissolution rate promotion, far too little attention has been paid to gather detailed documentations of the usage of the approach in enhancement of systemic drug efficiencies following drug bioavailability in human subjects or in suitable animal models. So as to affirm drug recrystallization as an efficient practice in intentionally increasing the bioavailability of poorly soluble drugs, further investigation in this field is mandatory. Ultimately, we could consider the potential management of crystal habits of poorly water soluble drugs as an approach for designing efficient pharmaceutical dosage forms.

3. Polymorphism

Very early on, in 1832 Wöhler and Liebig reported the first observation of polymorphism upon cooling a boiling solution of benzamide where needle-shaped crystals would initially formed followed by a conversion to rhombic crystals upon standing (WöhlerLiebig, 1832). After a longtime history, polymorphism has maintained its innovation for scientists as a curiosity and an urgent challenge of commercial relevance in manufacturing industries, as well.

In general, polymorphism is known to be the ability of a compound to crystallize in more than one distinct crystal structure. Evidently, structures of different crystals would receive various scopes of the possible ranges of intermolecular interactions like, van der Waals, ionic, and hydrogen bonds. It would not be far from expectations that the different polymorphs of the same molecule will have different free energies affecting all the basically physicochemical properties of the compound, in consequence. Therefore, the crystal structure of drugs have leading signature on both physical and chemical properties in a way that the solid-state forms might demonstrate variations in, color, stability, processability, solubility, dissolution and bioavailability, ranging from the subtle to the severe (Rodriguez-Spong et al., 2004). Despite the fact that distinct crystal habits have different processing issues or different chemical stability, these variations usually have direct solutions and the real impact of crystal forms is the difference in solubility and bioavailability. Figure 2 gives a fair picture of the probable consequences of a different crystal form in solubility. Decreased solubility of a certain crystalline form of a drug not only brings about lower bioavailability but also reduced rates of drug clearances are inevitable which are the inferences to the safety and efficacy of any drug product.

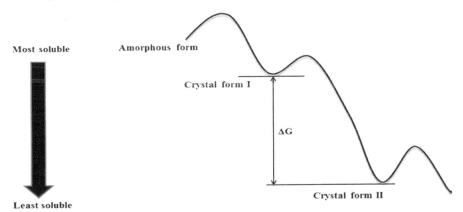

Fig. 2. Relative solubility of different crystal forms as a function of thermodynamic stability

Normally, producing the most thermodynamically stable polymorphic form of a drug in the range of interest is compulsory. The pharmaceutical science has been chiefly in charge for a change in this situation ever since the majority of drugs delivered orally receive rigid approval for a single crystal form or polymorph. Establishing the most thermodynamically stable form of a drug relies on obtaining comprehensive information about the existence and the interrelation of the polymorphs for a given active ingredient. Different polymorphs of a drug could be prepared by recrystallizing the drug with a range of solvents under the optimized conditions which has currently been arisen as an active research province of pharmaceutical science for improving the formulation related problems of drug molecules.

Accordingly, drug polymorphism investigations have an important role in any reformulation study since it has an impact on the development potential of a drug molecule, so as understanding the polymorphic tendencies of a drug molecule thorough characterization of the observed forms is of prime value.

The first important step in characterizing a polymorph of a drug is to distinguish between different structures of a molecule by its class. As it has been shown in figure 3, the most commonly observed forms in drug solids are the polymorphs, amorphous, crystals, solvates, and hydrates forms, that are fairly tractable from a processing notion. More to the point, hydrates or solvates are considered as pseudo-polymorphs that may either be an entirely different crystalline entity or simply incorporated in the parent crystal lattice; however, it is often possible to remove water or solvate by recrystallization.

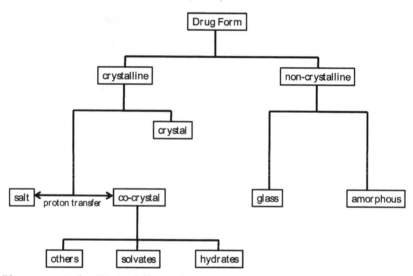

Fig. 3. Pharmaceutical solids in different forms.

While the molecular organization of drugs may differ having impact on the development potential of a drug molecule, the underlying concern for the drug performance is the same. Herein we tried to meet examples and detailed information on how recrystallization was used to improve the relative characteristics of drugs.

Park et al. designed a supercritical anti-solvent process for recrystallization of fluconazole, a triazole antifungal drug which has different polymorphic forms, to modify solid state characteristics of fluconazole by a range of operating conditions such as temperature, pressure, and type of solvent. Fluconazole particles were detected by means of differential scanning calorimetry, thermogravimetry analysis, powder X-ray diffraction, fourier transform infrared spectroscopy and scanning electron microscopy. Regarding the findings of the study, polymorphic forms of fluconazole were remarkably affected by the experimental conditions as the anhydrate form I of drug was obtained at low temperature and the anhydrate form II was obtained at higher temperatures. On the whole they suggested that the solid state characteristics of fluconazole, including the polymorphic form could effectively be managed by altering the operating conditions of the recrystallization process such as temperature, pressure, and solvent (Park et al., 2007). In another study, recrystallization of two anti-cancer drugs, erlotinib hydrochloride and fulvestrant was investigated employing the same method of recrystallization by Tien et al. Polymorph conversion from the original form B to form E or a prior form A for erlotinib was demonstrated under appropriate operational conditions which improved the dissolution rate of the drug. The micronized fulvestrant drug particles showed consistent polymorph like the original drug, but with differences in crystal habits. They also confirmed the positive effect of recrystallization in drug modifications (Tien et al., 2010).

Chen et al. evaluated the effects for the type of solvent, temperature and pressure as well as the solution flow rate on sulfathiazole particle formation using the same supercritical antisolvent technology for recrystallization. In the optimum operating condition of acetone as the solvent, temperature at 308 K with 12 MPa pressure and flow rate of 2mL/min the micronized sulfathiazole including polymorphic forms were obtained. Moreover, it was determined that various solvents resulted in different polymorphisms where the polymorph form III changed to form IV when ethanol was employed as the solvent. As a result, recrystallization of sulfathiazole produced in optimally micronized particles which exhibited a much narrower particle size distribution with an enhanced in vitro dissolution rate by 3.2 times to the original form of the drug (Chen et al., 2010).

4. Micro-crystal formation

Apart from the final quality of the drug crystal product in terms of purity, polymorphism, habit and morphology as well as crystal mechanical strength characterized in recrystallization procedure, size distributions of the drug particle is another crucial aspect of the pharmaceutical dosage forms. With reference to the Noyes–Whitney equation, application of a drug substance in a reduced particle size encourages bioavailability of the poorly water-soluble drug substances owing to the enhanced dissolution rate of micron- or nano-size drug particles (Chaumeil, 1998). An alternative area where small sized drug particles are indispensable is the pulmonary drug administration, inhaler drugs, in which drug powders should have a narrow particle size distribution and a mean particle size of 5 µm with almost no particles larger than 10 µm. Pharmaceutical dosage forms of these kinds are supposed to follow low particle agglomeration tendency, sufficient flow properties, and good batch-to-batch conformity in favor of the relative drug powders (IslamGladki, 2008).

Several techniques could be applied for the preparation of micron sized pharmaceuticals, such as mechanical comminution of the previously formed larger drug particles by crushing, grinding, milling and etc. that are the most common ways in this field. However, these methods not only provide limited opportunity for the control of important drug characteristics, like size, shape, morphology, surface properties but also ascertain distorted drug properties in a principally uncontrolled manner (RasenackMuller, 2004). As the surfaces in mechanically micronized drug powders are not naturally grown as the crystal cleaves at the crystal face and the surface energy changes, processing properties like flowability, agglomeration and stickiness to surfaces are the prevalent phenomenon. Employing milling processes such as jet milling, pearl-ball milling, or high-pressure homogenization other than influencing the preformulation behavior of drug structures, call for high energy and manpower being evidence for insufficient coverage of this method for application in reduced particle size drug production. It seems that these commonly being used techniques do not meet the ideal way for the production of small sized drug particles.

Since the suitable physicochemical and biopharmaceutical properties of a drug substance add to the time and cost of drug development, any postern to resolve these problems and produce small particles of drugs in a controlled process maximizes the opportunity to succeed in drug product manufacturing. Unlike the former techniques, production of small particles using controlled production processes such as spray drying, precipitation from supercritical fluid and recrystallization could be applied for the preparation of properly characterized micron sized pharmaceuticals. Microcrystallization in which the solubility and dissolution rate is improved by forming high specific surface area is used for preparation of drug microcrystals by recrystallization methods to reduce the size of the poorly water-soluble drug particles.

The usual technologies for recrystallization are fulfilled in this framework and it is to use solvent change or precipitation method by immediate mixing two liquids in presence of stabilizing agents. Regardless of the absolute efficiency of the recrystallization approach in production of large drug crystals, producing the small drug particles is still a sort of a challenge due to the high surface area of these particles, exerting tendency of a particle growth. So stabilizing agents would be foremost part in this system preventing particle growth by stabilizing the high specific surface area of small particles (Lechuga-BallesterosRodriguez-Hornedo, 1993). Nevertheless, recrystallization still continues to be one of the important parts of small sized drug production in drug development strategies. Therefore, microcrystals precipitate in presence of stabilizing protective polymers and a large and hydrophilized surface would be formed in a one process step having advantages over traditional milling techniques.

Hence, developing micro-crystallization as an efficient approach that modifies the biopharmaceutical and technological behavior of drug through selection of the process variables to reach an optimal pharmaceutical product has evolved to meet drug development challenges. Exploration on the growing number of publications in domain of micronized drug particle developing techniques apparently to enhance drug dissolution rate, considering the widely increasing number of poorly water-soluble drugs affirms the declaration.

Rasenack et al. prepared microcrystals of a poorly water-soluble drug ECU-01, an anti-inflammatory drug in preclinical state of development by a precipitation practice in the presence of stabilizing agents such as gelatin, chitosan, and different types of cellulose ethers and then spray-drying of the formed dispersion. Considering the low specific surface area of the nearly cuboid-like form of ECU-01, the aim of the survey was to enhance the drug dissolution rate by using microcrystals. Precipitation came off through dissolving the drug in acetone followed by an instant pouring an aqueous solution of the stabilizer into the drug solution. Via the use of cellulose type ethers as a stabilizer employed in this technique the thermodynamically unstable small particles were stabilized forming a protective layer on the crystal surface of the homogeneous microcrystals dispersions. Due to the polymorphic nature of the drug, the newly formed crystals appeared in a needle-shaped habit, highly increasing the specific surface area. Consequently, the dissolution rate rose up to 93% after 20 min compared to the 4% in common drug, indicating large surface of the microcrystals. Recrystallization of the poorly water-soluble drug ECU-01 was considered as a superior method which is easy to handle and only entails ordinary equipment (Rasenack et al., 2003).

A nonsteroidal anti-inflammatory drug, indomethacin, which has a hydrophobic and pH-dependent solubility nature, was developed and studied by Kim et al. through a microcrystallization technique to improve its physicochemical properties. Microcrystals of indomethacin was produced using a pH-shift procedure in which the drug was dissolved in an alkaline water to prepare saturated indomethacin solution thereafter the pH of the solution was decreased by adding 0.5N hydrochloric acid and stored at 20°C for 24 h to form microcrystal. The findings of the study exhibited similar physicochemical properties for the microcrystals produced and the standard crystalline powder in X-ray diffraction, differential scanning calorimetry, and Fourier transform infrared spectroscopy analyses, exclusive of a lower peak height in X-ray diffraction and somewhat lower melting temperature. The plate-like with uniform sized microcrystals of indomethacin dissolved about twice over the standard crystalline powder in the initial phase of dissolution study. Furthermore, the in vitro biological activity of the indomethacin microcrystals was assessed in their capacity to inhibit the proliferation of colon cancer cells that showed 20% greater activity than that of the standard crystalline powder. This view might have implications for improving the efficiency of chemotherapy in treating patients with malignant neoplasms using this technique for production of indomethacine microcrystals (Kim et al., 2003).

More recently, Talari et al. evaluated gliclazide microcrystals; a widely used drug for the treatment of non-insulin-dependent diabetes mellitus which shows a low solubility of 55 mg/L in water and gastric fluids leading to a low dissolution rate and variable bioavailability. The gliclazide microcrystals were prepared by in situ micronization techniques based on solvent and pH-shift and were examined for the drug absorption and pharmacokinetics of GL after oral administration in rats. Compared to the original drug, scanning electron microscopy showed significant changes in the shape and size of the prepared crystals using both methods. Recrystallized samples not only showed enhanced dissolution rates than untreated drug particles but also a reduced particle size of about 30 and 61 times by solvent-change and pH-shift methods were detected for drug crystals, respectively. Regarding results of the in vivo biological assays for hypoglycemic activity, microcrystallization of gliclazide using both methods resulted in an increased pharmacodynamic effect of glucose-lowering in diabetic rats which could be relevant to the improved dissolution rate of the drug (Talari et al., 2010).

Concisely, the microcrystallization of the drug particulates which has an effect on crystal habit, could also improve the drug absorption characteristics and the subsequent drug bioavailability.

5. Lyotropic liquid crystals formation

Liquid crystals (LCs) are a state of matter that has properties between those of a conventional liquid and those of a solid crystal. For instance, an LC may flow like a liquid, but its molecules may be oriented in a crystal-like way. There are many different types of LC phases, which can be distinguished by their different optical properties. When viewed under a microscope using a polarized light source, different liquid crystal phases will appear to have distinct textures. The contrasting areas in the textures correspond to domains where the LC molecules are oriented in different directions. Within a domain, however, the molecules are well ordered. LC materials may not always be in an LC phase (just as water may turn into ice or steam).

Liquid crystals can be divided into thermotropic, lyotropic and metallotropic phases. Thermotropic and lyotropic LCs consist of organic molecules. Thermotropic LCs exhibit a phase transition into the LC phase as temperature is changed. Lyotropic LCs exhibit phase transitions as a function of both temperature and concentration of the LC molecules in a solven . Metallotropic LCs are composed of both organic and inorganic molecules; their LC transition depends not only on temperature and concentration, but also on the inorganic-organic composition ratio.

Examples of liquid crystals can be found both in the natural world and in technological applications. Most modern electronic displays are liquid crystal based. Lyotropic liquid-crystalline phases are abundant in living systems. For example, many proteins and cell membranes are LCs. Other well-known LC examples are solutions of soap and various related detergents, as well as the tobacco mosaic virus.

A lyotropic liquid crystal consists of two or more components that exhibit liquid-crystalline properties in certain concentration ranges. In the lyotropic phases, solvent molecules fill the space around the compounds to provide fluidity to the system. In contrast to thermotropic liquid crystals, these lyotropics have another degree of freedom of concentration that enables them to induce a variety of different phases.

A compound, which has two immiscible hydrophilic and hydrophobic parts within the same molecule, is called an amphiphilic molecule. Many amphiphilic molecules show lyotropic liquid-crystalline phase sequences depending on the volume balances between the hydrophilic part and hydrophobic part. These structures are formed through the micro-phase segregation of two incompatible components on a nanometer scale. Soap is an everyday example of a lyotropic liquid crystal.

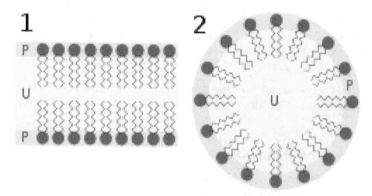

Fig. 4. Structure of lyotropic liquid crystal. The red heads of surfactant molecules are in contact with water, whereas the tails are immersed in oil (blue): bilayer (left) and micelle (right)

The content of water or other solvent molecules changes the self-assembled structures. At very low amphiphile concentration, the molecules will be dispersed randomly without any ordering. At slightly higher (but still low) concentration, amphiphilic molecules will spontaneously assemble into micelles or vesicles. This is done so as to 'hide' the hydrophobic tail of the amphiphile inside the micelle core, exposing a hydrophilic (water-soluble) surface to aqueous solution. These spherical objects do not order themselves in solution, however. At higher concentration, the assemblies will become ordered. A typical phase is a hexagonal columnar phase, where the amphiphiles form long cylinders (again with a hydrophilic surface) that arrange themselves into a roughly hexagonal lattice. This is called the middle soap phase. At still higher concentration, a lamellar phase (neat soap phase) may form, wherein extended sheets of amphiphiles are separated by thin layers of water. For some systems, a cubic (also called viscous isotropic) phase may exist between the hexagonal and lamellar phases, wherein spheres are formed that create a dense cubic lattice. These spheres may also be connected to one another, forming a bicontinuous cubic phase (VroegeLekkerkerker, 1992).

The objects created by amphiphiles are usually spherical (as in the case of micelles), but may also be disc-like (bicelles), rod-like, or biaxial (all three micelle axes are distinct). These anisotropic self-assembled nano-structures can then order themselves in much the same way as thermotropic liquid crystals do, forming large-scale versions of all the thermotropic phases (such as a nematic phase of rod-shaped micelles).

For some systems, at high concentrations, inverse phases are observed. That is, one may generate an inverse hexagonal columnar phase (columns of water encapsulated by amphiphiles) or an inverse micellar phase (a bulk liquid crystal sample with spherical water cavities).

A generic progression of phases, going from low to high amphiphile concentration, is: Discontinuous cubic phase (micellar cubic phase), Hexagonal phase (hexagonal columnar phase) (middle phase), Lamellar phase, Bicontinuous cubic phase, Reverse hexagonal columnar phase, and Inverse cubic phase (Inverse micellar phase)

Even within the same phases, their self-assembled structures are tunable by the concentration: for example, in lamellar phases, the layer distances increase with the solvent volume. Since lyotropic liquid crystals rely on a subtle balance of intermolecular interactions, it is more difficult to analyze their structures and properties than those of thermotropic liquid crystals. Similar phases and characteristics can be observed in immiscible diblock copolymers.

Lyotropic liquid crystals transitions occur with influence of solvents or recrystallization method. Lyotropic liquid crystals occur as a result of solvent induced aggregation of the constituent mesogens into micellar structure. This types of liquid crystalline states were mostly used for designing the sustain release drug delivery system. It is also used for improving the solubility and stability of insoluble drugs by incorporating it into micellar structure of liquid crystals (Lechuga-Ballesteros et al., 2003).

6. Spherical crystallization

Among the particles designed for solid pharmaceutical dosage forms, tablet supports for the half of all oral drug delivery system and 70% of the all pharmaceutical preparations produced. Direct tabletting, simple mixing and compressing drug powders have been widely applied to a large number of drugs on the industrial scale as an excellent technique. A successful tabletting and also compression of any drug material is reliant on the micromeritic properties of the drug crystals. In this context, crystals of needle-shaped or plated-shaped are challengeable to be handled with, due to the poor flowability of these crystals (Kaerger et al., 2004). In 1984 Kawashima et al. introduced crystal agglomeration with controlled properties as spherical crystallization to the pharmaceutical manufacturing and expressed that the spherically dense agglomerates were suitable for direct tabletting. Accordingly, crystallization and agglomeration of the drug substance particles concurrently in one step to transform crystals directly into compacted spherical form during the recrystallization process is defined as spherical crystallization (Kawashima, 1984). Spherical crystals could be established through two different techniques, either by typical spherical crystallization technique or non typical spherical crystallization technique (Nokhodchi et al., 2007). Non typical spherical crystallization technique might be regarded as the traditional crystallization process like salting-out, cooling, precipitation, whereas, the typical spherical crystallization is a three-solvent system employing three solvents; one is the drug dissolution medium known as the good solvent, another is a medium which partially dissolves the drug and has wetting feature that is named bridging liquid, and the last one is immiscible with the drug substance recognized as the bad solvent (Nokhodchi, Maghsoodi et al., 2007; Mahanty et al., 2010).

On average, spherical crystallization techniques are believed to be promising techniques in which the drug crystals are modified using different solvents for direct compressible spherical agglomerates, which can save money and time for tabletting. These approaches not only helps to achieve good flowability, compressibility and micromeritic properties of the drug substances, but also it is known to improve the the the wettability, bioavailability, and dissolution rate of some poorly soluble drugs (Kawashima, Handa et al., 1986; Di Martino et al., 1999; NokhodchiMaghsoodi, 2008). Besides, drug materials produced by the spherical crystallization technique result in the economical process in the development of the solid dosage forms for scaling up to a commercial level since it provides reduced time and cost by enabling faster operation, less machinery and fewer personnel.

Seeing that, magnesium aspartate and acetylsalicylic acid crystals in common are tetragonal and prism-shaped with different sizes they show poor flowability and compactibility properties indicating the crystal habit and the electrostatic charge. Szabo-Revesz et al. developed magnesium aspartate and acetylsalicylic acid via spherical crystallization since they are used in direct tablet-making and capsule-filling, the particle size and the spherical form are fundamental in view of their processibility. They prepared drug agglomerates through non-typical (magnesium aspartate) and typical (acetylsalicylic acid) spherical crystallization techniques. Crystal agglomerates of these drugs produced by these different spherical crystallization techniques created opportunity for a comparison between the results obtained. Considering the findings of the study, a higher initial cooling rate and a lower stirring rate were favorable in producing crystal agglomerates. The growth of particle size and the spherical form as well as the associated decreased specific surface of the magnesium aspartate and acetylsalicylic acid crystal agglomerates created better compactibility and cohesivity characteristics than the control samples in a softer flow time, and a higher bulk density. According to the authors both types of spherical crystallization (non-typical and typical) can be effectively used not only for spherical particle forming but also for size growing of drug materials that are pressed directly into tablets or made into filled capsules without excipients (Szabo-Revesz et al., 2002).

Celecoxib, a non-steroidal anti-inflammatory drug which is the first selective cyclooxygenase-2 inhibitor used in the treatment of osteoarthritis and rheumatoid arthritis exhibits poor flow and compression characteristics as well as incomplete and poor oral bioavailability due to its low aqueous solubility. Variety of papers in this case is representative of the suitability of this drug for spherical crystallization process to enhance the flow, compressibility and solubility properties of seems to be a beneficial objective to improve the following therapeutic efficacy of celecoxib (Banga et al., 2007). Paradkar et al. improved the micromeritic and compressional properties of celecoxib by a spherical crystallization process using the solvent change method. Selection of the solvents depended on the miscibility of solvents and also the solubility of drug in the solvents, which candidated acetone as the good solvent, dichloromethane as the bridging liquid, and water as the bad solvent. A solution of celecoxib in acetone was added to a solution of hydroxy propyl methylcellulose in dichloromethane. Drug was crystallized by adding the solution to wall-baffled vessel containing distilled water followed by a continuous mixing in a controlled speed style creating agglomerated spherical crystals. Celecoxib agglomerates exhibited satisfactory micromeritic, mechanical, and compressional properties demonstrating comparable in vitro drug release performance with the marketed capsule formulation (Paradkar et al., 2002). Elsewhere, Gupta et al. prepared spherical crystals of celecoxib using a more hydrophilic polymer, polyvinylpyrrolidone K-30 (PVP) and acetone, water and chloroform as solvent, non-solvent and bridging liquid, respectively. The agglomerates were determined by differential scanning calorimetry, X-ray diffraction, IR spectroscopic and scanning electron microscopy. They showed that the crystals possessed a good spherical shape with smooth and regular surface exhibiting significantly improved micromeritic properties compared to pure the drug. Moreover, the aqueous solubility and dissolution rate of the drug from crystals was notably increased nearly two times, with an increase in PVP concentration. In general, this technique may be applicable for producing oral solid dosage forms of other drugs with improved dissolution rate and oral bioavailability (Gupta et al., 2007). Correspondingly, use of these systems has the potential to facilitate drug development by saving valuable time in selecting the optimal physical or chemical characteristics of a given compound.

On the whole, lessening the risk of drug process modifications and providing the opportunity to gain more comprehensive rational property coverage would be established if such information is established through recrystallization processes at an early stage of drug developments. Eventually, it is of the utmost importance to strictly monitor the processing of drug substances with regards to the different crystal habits of drug materials, as well as to obtain a comprehensive understanding of the physical and chemical stability of these polymorphic states.

7. References

Adhiyaman, R. & Basu, S.K. (2006). Crystal modification of dipyridamole using different solvents and crystallization conditions. *International Journal of Pharmaceutics*, Vol.321, No.1-2, pp. 27-34, ISSN 03785173

Adibkia, K., Hamedeyazdan, S. & Javadzadeh, Y. (2011). Drug release kinetics and physicochemical characteristics of floating drug delivery systems. *Expert Opin Drug Deliv*, Vol.8, No.7, pp. 891-903, ISSN 1744-7593

Banga, S., Chawla, G., Varandani, D., Mehta, B.R. & Bansal, A.K. (2007). Modification of the crystal habit of celecoxib for improved processability. *J Pharm Pharmacol*, Vol.59, No.1, pp. 29-39, ISSN 0022-3573

Blagden, N., de Matas, M., Gavan, P.T. & York, P. (2007). Crystal engineering of active pharmaceutical ingredients to improve solubility and dissolution rates. *Adv Drug Deliv Rev*, Vol.59, No.7, pp. 617-630, ISSN 0169-409X

Carino, S.R., Sperry, D.C. & Hawley, M. (2006). Relative bioavailability estimation of carbamazepine crystal forms using an artificial stomach-duodenum model. *J Pharm Sci*, Vol.95, No.1, pp. 116-125, ISSN 0022-3549

Chaumeil, J.C. (1998). Micronization: a method of improving the bioavailability of poorly soluble drugs. *Methods Find Exp Clin Pharmacol*, Vol.20, No.3, pp. 211-215, ISSN 0379-0355

Chen, J. & Trout, B.L. (2008). Computational study of solvent effects on the molecular self-assembly of tetrolic acid in solution and implications for the polymorph formed from crystallization. *J Phys Chem B*, Vol.112, No.26, pp. 7794-7802, ISSN 1520-6106

Chen, Y.-M., Tang, M. & Chen, Y.-P. (2010). Recrystallization and micronization of sulfathiazole by applying the supercritical antisolvent technology. *Chemical Engineering Journal*, Vol.165, No.1, pp. 358-364, ISSN 13858947

Chow, A.H.L., Hsia, C.K., Gordon, J.D., Young, J.W.M. & Vargha-Butleff, E.I. (1995). Assessment of wettability and its relationship to the intrinsic dissolution rate of doped phenytoin crystals. *International Journal of Pharmaceutics*, Vol.126 pp. 21-28, ISSN 0378-5173

Dahlberg, C., Dvinskikh, S.V., Schuleit, M. & Furo, I. (2011). Polymer Swelling, Drug Mobilization and Drug Recrystallization in Hydrating Solid Dispersion Tablets Studied by Multinuclear NMR Microimaging and Spectroscopy. *Mol Pharm*, Vol.8, No.4, pp. 1247-1256, ISSN 1543-8392

Di Martino, P., Barthelemy, C., Piva, F., Joiris, E., Palmieri, G.F. & Martelli, S. (1999). Improved dissolution behavior of fenbufen by spherical crystallization. *Drug Dev Ind Pharm*, Vol.25, No.10, pp. 1073-1081, ISSN 0363-9045

Grzesiak, A.L., Lang, M., Kim, K. & Matzger, A.J. (2003). Comparison of the four anhydrous polymorphs of carbamazepine and the crystal structure of form I. *J Pharm Sci*, Vol.92, No.11, pp. 2260-2271, ISSN 0022-3549

Gupta, V.R., Mutalik, S., Patel, M.M. & Jani, G.K. (2007). Spherical crystals of celecoxib to improve solubility, dissolution rate and micromeritic properties. *Acta Pharm*, Vol.57, No.2, pp. 173-184, ISSN 1330-0075

Harbury, L. (1947). Solubility and melting point as functions of particle size; the induction period of crystallization. *J Phys Colloid Chem*, Vol.51, No.2, pp. 382-391, ISSN 0092-7023

Heng, J.Y., Bismarck, A. & Williams, D.R. (2006). Anisotropic surface chemistry of crystalline pharmaceutical solids. *AAPS PharmSciTech*, Vol.7, No.4, pp. 84, ISSN 1530-9932

Islam, N. & Gladki, E. (2008). Dry powder inhalers (DPIs)--a review of device reliability and innovation. *Int J Pharm*, Vol.360, No.1-2, pp. 1-11, ISSN 0378-5173

Jamali, F. & Mitchell, A.G. (1973). The recrystallization and dissolution of acetylsalicylic acid. *Acta Pharm Suec*, Vol.10, No.4, pp. 343-352, ISSN 0001-6675

Javadzadeh, Y., Mohammadi, A., Khoei, N.S. & Nokhodchi, A. (2009). Improvement of physicomechanical properties of carbamazepine by recrystallization at different pH values. *Acta Pharm*, Vol.59, No.2, pp. 187-197, ISSN 1330-0075

Jozwiakowski, M.J., Nguyen, N.A., Sisco, J.M. & Spancake, C.W. (1996). Solubility behavior of lamivudine crystal forms in recrystallization solvents. *J Pharm Sci*, Vol.85, No.2, pp. 193-199, ISSN 0022-3549

Kaerger, J.S., Edge, S. & Price, R. (2004). Influence of particle size and shape on flowability and compactibility of binary mixtures of paracetamol and microcrystalline cellulose. *Eur J Pharm Sci*, Vol.22, No.2-3, pp. 173-179, ISSN 0928-0987

Kawashima, Y. (1984). Development of spherical crystallization technique and its application to pharmaceutical systems. *Arch Pharm. Res.*, Vol.7, No.2, pp. 145-151

Kawashima, Y. (2003). Improved flowability and compactibility of spherically agglomerated crystals of ascorbic acid for direct tableting designed by spherical crystallization process. *Powder Technology*, Vol.130, No.1-3, pp. 283-289, ISSN 00325910

Kawashima, Y., Handa, T., Takeuchi, H., Okumura, M., Katou, H. & Nagata, O. (1986). Crystal modification of phenytoin with polyethylene glycol for improving mechanical strength, dissolution rate and bioavailability by a spherical crystallization technique. *Chem Pharm Bull (Tokyo)*, Vol.34, No.8, pp. 3376-3383, ISSN 0009-2363

Kim, S.T., Kwon, J.H., Lee, J.J. & Kim, C.W. (2003). Microcrystallization of indomethacin using a pH-shift method. *Int J Pharm*, Vol.263, No.1-2, pp. 141-150, ISSN 0378-5173

Kobayashi, Y., Ito, S., Itai, S. & Yamamoto, K. (2000). Physicochemical properties and bioavailability of carbamazepine polymorphs and dihydrate. *Int J Pharm*, Vol.193, No.2, pp. 137-146, ISSN 0378-5173

Krishnaiah S.R., Y. (2010). Pharmaceutical Technologies for Enhancing Oral Bioavailability of Poorly Soluble Drugs. *Journal of Bioequivalence & Bioavailability*, Vol.02, No.02, pp. 28-36, ISSN 0975-0851

Krishnaiah, Y.S.R. (2010). Pharmaceutical technologies for enhancing oral bioavailability of poorly soluble drugs. *Journal of Bioequivalence & Bioavailability*, Vol.2, No.2, pp. 028-036, ISSN 0975-0851

Lechuga-Ballesteros, D., Abdul-Fattah, A., Stevenson, C.L. & Bennett, D.B. (2003). Properties and stability of a liquid crystal form of cyclosporine-the first reported naturally occurring peptide that exists as a thermotropic liquid crystal. *J Pharm Sci*, Vol.92, No.9, pp. 1821-1831, ISSN 0022-3549

Lechuga-Ballesteros, D. & Rodriguez-Hornedo, N. (1993). Growth and morphology of L-alanine crystals: influence of additive adsorption. *Pharm Res*, Vol.10, No.7, pp. 1008-1014, ISSN 0724-8741

Liebenberg, W., de Villiers, M.M., Wurster, D.E., Swanepoel, E., Dekker, T.G. & Lotter, A.P. (1999). The effect of polymorphism on powder compaction and dissolution properties of chemically equivalent oxytetracycline hydrochloride powders. *Drug Dev Ind Pharm*, Vol.25, No.9, pp. 1027-1033, ISSN 0363-9045

Llinas, A., Burley, J.C., Box, K.J., Glen, R.C. & Goodman, J.M. (2007). Diclofenac solubility: independent determination of the intrinsic solubility of three crystal forms. *J Med Chem*, Vol.50, No.5, pp. 979-983, ISSN 0022-2623

Lobenberg, R. & Amidon, G.L. (2000). Modern bioavailability, bioequivalence and biopharmaceutics classification system. New scientific approaches to international regulatory standards. *Eur J Pharm Biopharm*, Vol.50, No.1, pp. 3-12, ISSN 0939-6411

Maghsoodi, M., Hassan-Zadeh, D., Barzegar-Jalali, M., Nokhodchi, A. & Martin, G. (2007). Improved compaction and packing properties of naproxen agglomerated crystals obtained by spherical crystallization technique. *Drug Dev Ind Pharm*, Vol.33, No.11, pp. 1216-1224, ISSN 0363-9045

Mahanty, S., Sruti, J., Niranjan Patra, C. & Bhanoji Rao, M.E. (2010). Particle design of drugs by spherical crystallization techniques. *International Journal of Pharmaceutical Sciences and Nanotechnology*, Vol.3, No.2, pp. 912-918, ISSN 0974 – 9446

Mirmehrabi, M. & Rohani, S. (2005). An approach to solvent screening for crystallization of polymorphic pharmaceuticals and fine chemicals. *J Pharm Sci*, Vol.94, No.7, pp. 1560-1576, ISSN 0022-3549

Nokhodchi, A., Bolourtchian, N. & Dinarvand, R. (2003). Crystal modification of phenytoin using different solvents and crystallization conditions. *Int J Pharm*, Vol.250, No.1, pp. 85-97, ISSN 0378-5173

Nokhodchi, A. & Maghsoodi, M. (2008). Preparation of spherical crystal agglomerates of naproxen containing disintegrant for direct tablet making by spherical crystallization technique. *AAPS PharmSciTech*, Vol.9, No.1, pp. 54-59, ISSN 1530-9932

Nokhodchi, A., Maghsoodi, M. & Hassanzadeh, D. (2007). An Improvement of Physicomechanical Properties of Carbamazepine Crystals. *Iranian Journal of Pharmaceutical Research*, Vol.6, No.2, pp. 83-89, ISSN 1735-0328

Paradkar, A.R., Pawar, A.P., Chordiya, J.K., Patil, V.B. & Ketkar, A.R. (2002). Spherical crystallization of celecoxib. *Drug Dev Ind Pharm*, Vol.28, No.10, pp. 1213-1220, ISSN 0363-9045

Park, H.J., Kim, M.S., Lee, S., Kim, J.S., Woo, J.S., Park, J.S. & Hwang, S.J. (2007). Recrystallization of fluconazole using the supercritical antisolvent (SAS) process. *Int J Pharm*, Vol.328, No.2, pp. 152-160, ISSN 0378-5173

Perlovich, G.L., Strakhova, N.N., Kazachenko, V.P., Volkova, T.V., Tkachev, V.V., Schaper, K.J. & Raevsky, O.A. (2008). Sulfonamides as a subject to study molecular interactions in crystals and solutions: sublimation, solubility, solvation, distribution and crystal structure. *Int J Pharm*, Vol.349, No.1-2, pp. 300-313, ISSN 0378-5173

Rasenack, N., Hartenhauer, H. & Muller, B.W. (2003). Microcrystals for dissolution rate enhancement of poorly water-soluble drugs. *Int J Pharm*, Vol.254, No.2, pp. 137-145, ISSN 0378-5173

Rasenack, N. & Muller, B.W. (2004). Micron-size drug particles: common and novel micronization techniques. *Pharm Dev Technol*, Vol.9, No.1, pp. 1-13, ISSN 1083-7450

Rodriguez-Spong, B., Price, C.P., Jayasankar, A., Matzger, A.J. & Rodriguez-Hornedo, N. (2004). General principles of pharmaceutical solid polymorphism: a supramolecular perspective. *Adv Drug Deliv Rev*, Vol.56, No.3, pp. 241-274, ISSN 0169-409X

Seton, L., Roberts, M. & Ur-Rehman, F. (2010). Compaction of recrystallised ibuprofen. *Chemical Engineering Journal*, Vol.164, No.2-3, pp. 449-452, ISSN 1385-8947

Sinclair, W., Leane, M., Clarke, G., Dennis, A., Tobyn, M. & Timmins, P. (2011). Physical stability and recrystallization kinetics of amorphous ibipinabant drug product by fourier transform raman spectroscopy. *J Pharm Sci*, pp., ISSN 1520-6017

Suzuki, T., Araki, T., Kitaoka, H. & Terada, K. (2010). Studies on mechanism of thermal crystal transformation of sitafloxacin hydrates through melting and recrystallization, yielding different anhydrates depending on initial crystalline forms. *Int J Pharm*, Vol.402, No.1-2, pp. 110-116, ISSN 1873-3476

Szabo-Revesz, P., Hasznos-Nezdei, M., Farkas, B., Goczo, H., Pintye-Hodi, K. & Eros, I. (2002). Crystal growth of drug materials by spherical crystallization. *Journal of Crystal Growth*, Vol.237, No.239, pp. 2240–2245, ISSN 0022-0248

Talari, R., Varshosaz, J., Mostafavi, S.A. & Nokhodchi, A. (2010). Gliclazide microcrystals prepared by two methods of in situ micronization: pharmacokinetic studies in diabetic and normal rats. *AAPS PharmSciTech*, Vol.11, No.2, pp. 786-792, ISSN 1530-9932

Tian, F., Sandler, N., Aaltonen, J., Lang, C., Saville, D.J., Gordon, K.C., Strachan, C.J., Rantanen, J. & Rades, T. (2007). Influence of polymorphic form, morphology, and excipient interactions on the dissolution of carbamazepine compacts. *J Pharm Sci*, Vol.96, No.3, pp. 584-594, ISSN 0022-3549

Tien, Y.-C., Su, C.-S., Lien, L.-H. & Chen, Y.-P. (2010). Recrystallization of erlotinib hydrochloride and fulvestrant using supercritical antisolvent process. *The Journal of Supercritical Fluids*, Vol.55, No.1, pp. 292-299, ISSN 0896-8446

Tiwary, A.K. (2001). Modification of crystal habit and its role in dosage form performance. *Drug Dev Ind Pharm*, Vol.27, No.7, pp. 699-709, ISSN 0363-9045

Vroege, G.J. & Lekkerkerker, H.N.W. (1992). Phase transitions in lyotropic colloidal and polymer liquid crystals. *Rep. Fmg. Phys.*, Vol.55, pp. 1241-1309, ISSN 0034-4885

Wagner, D., Glube, N., Berntsen, N., Tremel, W. & Langguth, P. (2003). Different dissolution media lead to different crystal structures of talinolol with impact on its dissolution and solubility. *Drug Dev Ind Pharm*, Vol.29, No.8, pp. 891-902, ISSN 0363-9045

Wöhler, F. & Liebig, J. (1832). Untersuchungen über das Radikal der Benzoesäure. *Annalen der Pharmacie*, Vol.3, No.3, pp. 249-282, ISSN 0075-4617

Yeo, S., Kim, M. & Lee, J. (2003). Recrystallization of sulfathiazole and chlorpropamide using the supercritical fluid antisolvent process. *The Journal of Supercritical Fluids*, Vol.25, No.2, pp. 143-154, ISSN 0896-8446

9

Recrystallization of Enantiomers from Conglomerates

Valérie Dupray
Université de Rouen
France

1. Introduction

An object is considered as "chiral" if it is not superimposable on its mirror image. Due to the presence of asymmetric carbon atoms, numerous molecules are chiral but other stereogenic centers provide the asymmetric character of molecular compound (e.g. sulfoxides). Metal-organic complexes can also be asymmetric even if the ligands do not have any stereogenic center. The equimolar mixture of two enantiomers is the racemic mixture. Except for their rotatory powers which are equal in value but of opposite signs, enantiomers present identical chemical and physical properties in achiral environment. Conversely, their behaviors in chiral environment such as biological systems are often different. In particular, for chiral drugs, the pharmacological and the toxicological effects can be drastically dissimilar.

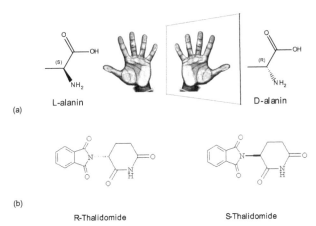

(a) L-alanin   D-alanin

(b) R-Thalidomide   S-Thalidomide

Fig. 1. Enantiomers of (a) Alanin and (b) Thalidomide

A well-known example is the case of thalidomide. Thalidomide was a drug prescribed for nausea during pregnancy in the later 1950s and was marketed as the racemic mixture. The R-enantiomer showed the desired curing effect while the S-enantiomer caused serious teratogenic effects (Kocher-Becker et al, 1982). The health scandal that follows led to the removal of the drug in 1961. Since, the pharmaceutical regulatory authorities have increased the pressure for chiral drugs to be administered in an enantiomerically pure form. Thus, chiral drug industry has been in constant development and has become the most growing segment of the drug market. It represents now more than one third of the drug sales worldwide (Stinson, 2000). At the same time, chemical producers have developed new enantiomeric intermediates for industry and especially enantioselective technologies. Asymmetric synthesis (a chemical reaction of an enantiomeric agent or catalyst with a substrate to produce a single enantiomer of the desired molecule) remains the first mode of production of pure enantiomers. However, the reaction product can appear to be a racemic mixture due to synthesis conditions or in rare occasions to racemization. Commonly, it is also cheaper, easier or cleaner to synthesize directly the racemic mixture. In those cases, a chiral resolution (separation) has to be performed to recover the pure enantiomer.

A direct enantiomeric separation can be carried out via chiral chromatography. To this purpose, the inner surface of a chromatographic column is bonded or coated with a chiral selector or alternatively the chiral selector can be incorporated directly to the stationary phase. As a result, the two enantiomers have different retention times which allow the resolution (Subramanian, 2006). However, this process which leads to high purity products, the productivity (which can be poor due to a high separation time and/or a small rate of injection of the product in the column) is not always compatible with industrial standards. Another solution consists of using crystallization.

The most popular crystallization method certainly remains the Pasteurian resolution (Pasteur, 1853) for which a chiral resolving agent is used to obtain the crystallization of diastereoisomers. Typically, the resolving agent (an enantiomerically pure acid (resp. base)) is added to the racemic base (resp. acid) to form diastereoisomeric salts. Contrary to the enantiomers, diastereoisomers do not exhibit the same symmetry (Coquerel, 2000) and as a consequence do not present the same chemical and physical properties (in particular solubility). Dissolution-recrystallization in an appropriate solvent or mix of solvents permits to obtain a single solid crystalline phase (the pure salt). Then, the pure enantiomer can be recovered by salting out to remove the resolving agent.

In order to improve the yield and to reduce the quantity of resolving agents, several variations have been proposed. Let us just mention the Marckwald's method (Marckwald, 1896), the Pope and Peachey's method (Pope & Peachey, 1899) and more recently the Dutch resolution (Vries et al, 1998), (Kaptein et al, 2000). Note also that the accurate determination of the phase diagrams under the conditions of the experimental process allows an optimization of the resolution (Marchand et al, 2004). Even imperfect (a yield close to 100% is rarely obtained), this method most often fits industrial and laboratories requirements.

Besides the Pasteurian method, the resolution can be performed by "preferential crystallization" (PC) also called "crystallization by entrainment". PC is a stereoselective process in which, alternatively, for a given period of time, only one enantiomer crystallizes although both enantiomers are supersaturated in the mother liquor. In 1866, Gernez (Gernez, 1866) was the first to observe that a saturated solution of one enantiomer, seeded by the same enantiomer, allows the formation of enantiomerically pure crystals. Conversely, he noted that if the seeding was done with the other enantiomer, no crystallization was observed. The entrainment phenomenon itself was described by Jungfleish (Jungfleish, 1882) who underlines the predominant influence of supersaturation.

In the 1990s, the phenomena involved in the preferential crystallization have been explained in details in literature using phase diagrams. For a complete description please refer to Jacques and his co-workers book "Enantiomers, Racemates, and Resolution"(Jacques et al, 1994).

Several improvements of this cyclic process have been proposed such as the auto-seeded variant AS3PC developed by Professor Gerard Coquerel and co-workers at the French University of Rouen (Ndzié et al, 1997; Coquerel, 2007). AS3PC by its robustness, its low cost and reproducibility offers real possibilities for industrial applications of preferential crystallization. As a proof, the AS3PC process has been the subject of several patents for the University of Rouen (Coquerel et al, 1994; Coquerel et al, 1995; Helmreich et al, 2010)

The main advantage of PC and its derivatives is certainly that they do not require any resolving agent. However they suffer from a serious limitation: the compound to be resolved or accessible derivatives such as salts solvated or not, must crystallize as a conglomerate (ie. an equimolar mechanical mixture of crystals, each one containing only a single enantiomer). The requirement of a conglomerate is an important restriction to the application of preferential crystallization due to the low occurrence of conglomerates among the molecular crystallized compounds (5-10% of the racemic species only)(Jacques et al, 1994).

Because the detection of a conglomerate is a key step of the resolution process, we chose to focus this contribution on this subject. The first part constitutes a necessary reminder about crystal packing of chiral molecules and includes the definition of the different types of structures that can be encountered crystallizing racemic mixtures. The second part is dedicated to the enantiopurification by crystallization. The benefits of working with conglomerates are underlined. The next part describes the detection of conglomerates itself. Classical methods are first recalled and then the prescreening of a conglomerate via a high throughput technique involving nonlinear optics is presented. In the last part, we propose a sequential diagram to optimize the detection of conglomerates.

2. Crystal structure and packing of chiral molecules

Speaking about crystallization and recrystallization implies to take into account the way the molecules are packed inside the crystal. It is all the more important as the chiral nature of a molecule imposes some limitations.

2.1 Types of packing of chiral molecules

Three most common types of packing are usually observed when crystallizing a racemic mixture (Coquerel, 2000):

- The racemic compound is the most common (90-95% of racemic species). In the vast majority of the cases it is a <1-1> stoichiometric compound.
- The conglomerate is a mechanical mixture of single crystals containing homochiral molecules only. As mentioned previously, conglomerates represent only 5 to 10% of the racemic species (Jacques et al, 1994).
- The racemic solid solution has a low occurrence. It is a solid solution containing an equal number of molecules of each enantiomer but contrary to the racemic compound, the arrangement is a random distribution (see fig.2)

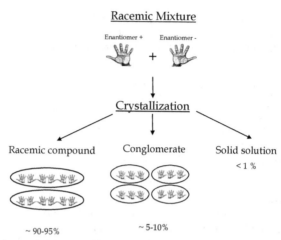

Fig. 2. Types of packing of chiral molecules

2.2 Chirality of crystalline structures

The nature of the crystalline structure formed by the chiral molecules is of major importance when considering the recrystallization processes. At this stage, it is imperative to distinguish between the chirality of the molecule itself and the chirality of the crystalline structure. Indeed, crystalline structures formed from enantiomers (chiral molecules) present in racemic mixture are not necessarily chiral. Let us just first recall that chiral crystals belong to space groups that contain only symmetry operation of the first kind (rotation, translation). It excludes the symmetry operations of the second kind (rotoinversion) which are allowed in non-centrosymmetric structures.

Crystalline structures can then be classified into three categories (Flack, 2003; Galland et al, 2009):

- Centrosymmetric (achiral) structures (type CA) which corresponds to point groups $\bar{1}$, 2/m, mmm, 4/m, 4/mmm, $\bar{3}$, $\bar{3}$m, 6/m, 6/mmm, m3 and m3m.
- Non centrosymmetric achiral structures (type NA) for point groups m, mm2, $\bar{4}$, 4mm, $\bar{4}$ 2m, 3m, $\bar{6}$, 6mm, $\bar{6}$ m2 and $\bar{4}$ 3m.
- Non centrosymmetric chiral structures (type NC) associated with point groups 1, 2, 222, 4, 422, 3, 32, 6, 622, 23 and 432.

Racemic compounds can theoretically be part of any space groups (CA, NA or NC) but 95 % of the known racemic compounds crystallize in centrosymmetric space groups (CA). The predominant space groups are: $P2_1/c$, $C2/c$, Pbca and $P\bar{1}$ (95 %) of the centrosymmetric racemic compounds)(Dalhus et al, 2000). Non centrosymmetric racemic compounds (NA) like DL-Allyglycine (Dalhus et al, 2000) represent a proportion of 4.5-5 %, mainly placed in space groups Pna21, Pca21, Cc and Pc. Rare cases of racemic compounds crystallizing in chiral space groups (mainly $P2_12_12_1$ and $P2_1$) have been reported (ortho-thyrosine, α-methylsuccinic acid or camphoroxime (Jacques et al, 1994; Brock et al, 1991; Kostaniovsky, 2008). Their occurrence is estimated to be only 0.02%(Flack et al, 2003).

Regarding the conglomerate, the chiral nature of the molecules imposes restriction in the construction of the crystalline structure so it is impossible to form achiral crystal structures by crystallization of enantiomerically pure chiral molecules (Jacques et al, 1994), (Flack et al, 2003). Consequently, conglomerates crystallize only in one of the 65 chiral space groups (spaces groups $P2_12_12_1$, $P2_1$, C_2 and P_1 represent 95 % of the known conglomerates (Belsky et, 1977).

		Achiral structure		Chiral structure
		CA Centrosymmetric Achiral	**NA** Non centrosym. Achiral	**NC** Non centrosym. Chiral
Racemic compound 90 – 95 %	*Structure*	Permitted	Permitted	Permitted
	Proportion	~ 95 %	4.5-5 %	0.02%
	Predominant space groups	$P2_1/c$, C_2/c, Pbca and P$\bar 1$	$Pna2_1$, $Pca2_1$, Cc and Pc	$P2_12_12_1$ and $P2_1$
Conglomerate 5-10%	*Structure*	Forbidden	Forbidden	Mandatory
	Proportion	0%	0%	100%
	Predominant space groups	/////////	/////////	$P2_12_12_1$, $P2_1$, C2 and P1
Solid Solution Less than 1%	*Structure*	Permitted	Permitted	Permitted

Table 1. Formation of crystalline structures from racemic mixtures

Concerning the solid solution, the three crystal types are permitted but no data seems available about the predominant space groups in the case of racemic solid solution. These various possibilities for the crystallization of racemic mixtures are summarized in Table 1.

3. Recrystallization of enantiomers

Enantiopurification by crystallization is presented here for two cases:

-   a racemic compound-forming system using the classical selective crystallization processes
-   a conglomerate for which a crystallization by entrainment is possible

The two recrystallization procedures require a good knowledge of the behavior of the two enantiomers regarding melting (binary phase diagrams) and solubility in a given solvent (Ternary phase diagrams). For a detail description of the various phase diagrams encountered for these systems of enantiomers, please refer to (Jacques et al, 1994), (Lorenz et al, 2006) and (Coquerel, 2000).

3.1 Enantiopurification from racemic compound-forming system

On figure 3 is depicted the usual ternary phase diagram observed for a racemic compound-forming system (S and R enantiomers, racemic compound RS and solvent V_1) at a given temperature T_1. Six regions are delimited by the solubility curves (see legend of figure 3) and can contain from 1 to 3 phases. The size of theses different regions varies of course with temperature.

Assuming that the mixture of enantiomers to purify (represented by point M) has an enantiomeric excess exceeding that of the invariant liquid I_1, the best process of enantiopurification consists in adding solvent V_1 in order to place the overall composition into the two-phase region (region 3 if enantiomer S is concerned). A new equilibrium is then established between <S> and the mother liquor (point K). Then, a selective crystallization using an appropriate cooling program can be used to obtain the desired enantiomer.

Depending on the location of I_1 and K, the maximum mass of pure enantiomer <S> that can be recovered is given by:

$$m<S> = m_T \times \frac{KI_1}{SI_1} \tag{1}$$

with m_T, the total mass.

Obviously, this mass is always lower than the total mass of enantiomeric excess since e.e.$(I_1) \neq 0$.

It is then possible to proceed to a recrystallization of a pure enantiomer in the case of a racemic compound-forming system. However, this is only applicable to a starting system presenting a sufficient enantiomeric excess. Thus, the process requires a first enrichment step that can be performed by chiral chromatography or via diastereoisomer formation. Moreover, it is not possible to recover the totality of the enantiomeric excess.

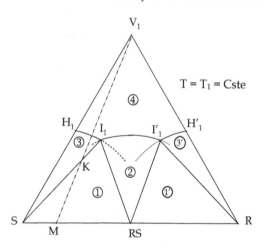

Triphasic domains
　　　Region 1 : Invariant liquid I_1 doubly saturated + <S> + <RS>
　　　Region 1': Invariant liquid I'_1 doubly saturated + <R> + <RS>
Biphasic domains:
　　　Region 2: Saturated solution + <RS>
　　　Region 3: Saturated solution + <S>
　　　Region 3': Saturated solution + <R>
Monophasic domain:
　　　Region 4: Undersaturated solution

Fig. 3. Ternary phase diagram observed for a racemic compound-forming system

3.2 Enantiopurification from a conglomerate-forming system

On figure 4 is depicted the ternary phase diagram observed for a conglomerate forming system (S and R enantiomers, solvent V_1) at temperature T_1. This diagram is simpler that the one corresponding to the racemic compound-forming system since only two solid phases (S and R) can coexist. Four regions are delimited by the solubility curves (see legend of figure 4).

Here, as for the racemic compound, starting from a mixture M of enantiomers, the process consists in adding the exact quantity of solvent (at T_1) so that the overall synthetic mixture is represented by point K. Point K is situated on the tie-line SI separating the 3 phase domain I-S-R (which contains: pure <S>, pure <R> and the doubly saturated solution I) and the biphasic domain S-H-I (where the phases in equilibrium are pure <S> and the saturated solution). Then the selective crystallization can be pursued.

Contrary to the previous case, the whole enantiomeric excess of mixture M can be retrieved (on the condition that the compound crystallizes as a conglomerate without partial solid solution). It constitutes the best situation to obtain an efficient recrystallization in thermodynamic equilibrium.

In practice (see figure 5), it is better to heat the system above T_1 so that the suspension at $T_2 > T_1$ contains only pure <S> and a saturated solution. The smooth cooling program will mainly consist in crystal growth leading to a better filterability and a more efficient washing of the filtration cake.

It is even sometime possible to retrieve more than the initial enantiomeric excess by cooling the system at T_0. Point K is then situated in the 3-phase domain and as long as the crystallization is stereoselective the liquid composition can evolve from I_1 at (T_1) to Z_0 at T_0 (figure 6). Note that Z_0 is located on the metastable solubility curve of S at T_0. This corresponds to an "entrainment" (i.e. a single operation of the so-called AS3PC process).

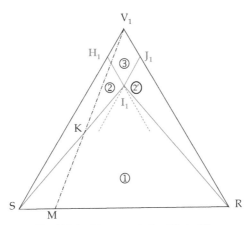

Triphasic domain: 1 - Invariant liquid I doubly saturated + <S> + <R>
Biphasic domains: 2 - Saturated solution + <S>; 2' - Saturated solution + <R>
Monophasic domain: 3 - Undersaturated solution
Dashed lines stand for metastable solubilities

Fig. 4. Ternary phase diagram observed for a conglomerate forming system

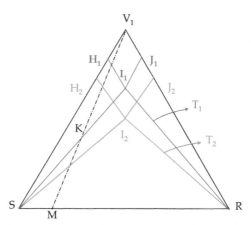

Fig. 5. Ternary phase diagrams observed for a conglomerate-forming system at T_1 (Blue) and T_2 (Red)

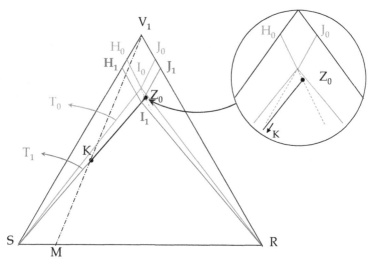

Fig. 6. Principle of crystallization by entrainment

When the initial mixture M corresponds already to a high enantiomeric excess, solvent V_1 at temperature T_1 won't be appropriate because the slurry will be simply much too viscous. Thus, it is necessary to find another solvent V_2 in which the pure enantiomer <S> is poorly soluble (figure 7).

3.3 Enantiopurification from conglomerate-forming derivatives

Figure 8 shows that a conglomerate-forming solvate (here a monohydrate) can also give the full discrimination in the solid state even if a stable racemic compound exists for the anhydrous chiral components. The enantiopurification by means of crystallization can be operated in a similar way as that presented above (point K for a mixture M at T_1). When the initial point M is close to the pure enantiomer (e.g. S), the medium can be changed by mixing an anti-solvent. The second solvent must be miscible with water, should not induce a miscibility gap and should not inhibit the formation of the conglomerate.

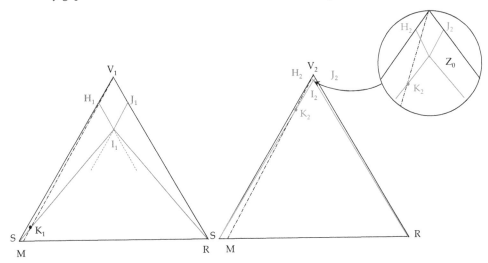

Fig. 7. Enantiomeric purification with a high enantiomeric excess – Choice of another solvent

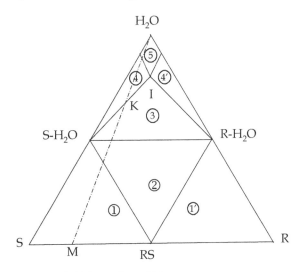

1 : <S-H$_2$O> + <S> + <RS> ; 1': <R-H$_2$O> + <S> + <RS>
2: <S-H$_2$O> + <R-H$_2$O> + <RS>
3: Invariant liquid I doubly saturated + <S-H$_2$O> + <R-H$_2$O>
4: Saturated solution + <S-H$_2$O>
4': Saturated solution + <R-H$_2$O>
5: Undersaturated solution

Fig. 8. Enantiomeric purification from a derivative (case of a monohydrate)

3.4 Benefits of conglomerate

It is now obvious that proceeding to the enantiopurification with a starting product crystallizing as a conglomerate is even more beneficial since:

- an optimized recrystallization can 'virtually' be carried out without any loss of enantiomeric excess which is not possible with a racemic compound.
- the choice of appropriate solvent and crystallization conditions usually allows to reach an e.e greater than 99.5 %.
- an optimized recrystallization can be carried out on derivatives such as hydrates (crystallizing as conglomerates) even if the molecule of interest crystallizes as a stable racemic compound.

Note also, that when coupled to racemization, almost pure enantiomer can be obtained by attrition from the racemic mixture (Levillain et al, 2009).

4. Detection of conglomerate

To perform preferential crystallization, the racemic mixture should crystallize as a stable conglomerate. In most cases, the chiral resolution can not be envisaged on the original compound because it does not fulfill this condition. An interesting alternative is to proceed to the resolution on derivatives such as salts, co-crystals, hydrates or solvates that crystallize as conglomerates (at the end of the resolution process, the pure enantiomer can be easily recovered by salting out, dehydration or desolvation). To multiply the chances of spotting a conglomerate, experiments have to be performed on a large number of non chiral acids (for a chiral base) with various stoichiometries and solvents, with different co-crystal formers and under various crystallization conditions.

Consequently, starting from the racemic mixture it is often necessary to synthesize series of derivatives that have to be analyzed in order to isolate at least one conglomerate forming system. Due to the low occurrence of conglomerates, it usually implies to investigate a large number of derivatives.

Conglomerate detection which is an essential step of the resolution method can be pursued by various methods which are described and discussed in the following.

4.1 Classical screening of conglomerates

4.1.1 Comparisons between IR, Raman, solid state NMR or XRPD patterns

Several spectroscopic techniques can be used to differentiate the racemic mixture and the conglomerate via comparison between the racemic pattern and that of the enantiomer. In case of conglomerate without partial solid solution, the patterns of the pure enantiomer obtained from IR, Raman, solid state NMR or XRPD should be superimposable to the ones obtained for the racemic mixture.

Even if Raman spectroscopy and IR spectroscopy can be of a certain help in spotting a conglomerate, they cannot be considered as totally reliable. Indeed, vibrational bands mainly arise from molecular vibrations. Only low frequency vibrations are directly related to the vibrations of the crystal lattice. If the neighborhood of a given molecule can generate variations in the high frequency domain, two close structures can generate similar spectra. Considering these elements, the perfect match of the XRPD patterns is surely the most dependable way to conclude on the conglomerate nature of a sample. XRPD is of particular interest as it also permits to check the crystallinity of the sample and to spot partial or total solid solutions (Wermester et al, 2007; Renou et al, 2007). Moreover, if a single crystal of sufficient size is available, a structure can be resolved by single crystal X-Ray diffraction. The knowledge of the crystal structure definitively confirms the conglomerate nature (or not) of the sample.

4.1.2 Alternatives methods

Other methods consist of isolating a unique particle (single crystal) from the racemic mixture to proceed to the analysis. This particle can be dissolved in a nematic phase (Jacques et al, 1994) or analyzed by chromatography (Chiral CG, HPLC) (Pirzada et al, 2010) and/or polarimetry. However, there are serious limitations to these techniques since the collection of an isolated particle can be difficult and also because the analysis of a single particle necessitates high detection levels. To be exhaustive, let us also mentioned that the detection of a conglomerate can also be performed via thermal analysis (the construction of the binary phase diagram by simple measurements of the melting temperatures of the enantiomers and that of their corresponding racemic mixture can be used to identify the nature of the sample (conglomerate or racemic compound)(Neau et al, 1993) or Solid state circular dichroïsm (SS-CD) and CD microscopy (Kuroda et al, 2000; Claborn et al, 2003). The construction of a ternary isotherm is also a reliable method especially when a conglomerate of solvated phases is suspected.

4.2 High throughput technique for conglomerate prescreening

Unfortunately and despite technical improvements, all the previous methods are time consuming and often require the pure enantiomer to be available at an early stage of the process. Consequently, there is a need for a faster (and cheaper) method compatible with a combinatorial approach of crystallization derivatives (salts, hydrates, solvates, cocrystals, etc.)

Considering the data summarized in table 1, it appears that racemic compounds crystallize not often in non centrosymmetric structures (less than 5% for NA + NC). Consequently, a method that would detect the absence of center of symmetry in crystals obtained from racemic mixtures will be able in most cases to detect conglomerates. To this purpose, nonlinear optics can be very useful. In the next paragraphs are presented the theory of second harmonic generation in crystals and described the principle of conglomerate prescreening.

4.2.1 Nonlinear optics and second harmonic generation (SHG)

In a given medium, the propagation of light is mainly driven by the dielectric properties and the response to electromagnetic fields. The application of an electromagnetic field (light) to a molecule modifies the shape of the electronic cloud and consequently creates an induced electric dipole moment.

The term polarization (**P**, vectorial dipole moment per unit volume) is used to describe this phenomenon on a macroscopic scale:

$$\mathbf{P} = \varepsilon_0 \, \chi^{(1)} \mathbf{E} \tag{2}$$

with **E**, the applied electric field, ε_0 is the vacuum permittivity and $\chi^{(1)}$ the linear (first order) susceptibility of the material.

The linear susceptibility is a 2nd rank tensor related to the permittivity ε and the refractive index n of the material:

$$\varepsilon = n^2 = \varepsilon_0 \left(1 + \chi^{(1)}\right) \tag{3}$$

For week electric fields, the polarization varies linearly with **E**. However, for high light intensities (typically greater than 1 MW/cm^2), the polarization becomes a nonlinear function of the applied electric field. It can be expressed as a power series expansion of the macroscopic field:

$$\mathbf{P} = \varepsilon_0 \left(\chi^{(1)} \, \mathbf{E} + \chi^{(2)} \, \mathbf{E}^2 + \chi^{(3)} \, \mathbf{E}^3 + \ldots\right) = \varepsilon_0 \chi^{(1)} \, \mathbf{E} + \mathbf{P}_{NL} \tag{4}$$

with $\chi^{(2)}$ and $\chi^{(3)}$ the second and the third order susceptibility tensors respectively and $\mathbf{P}_{NL}$, the nonlinear polarization. This equation implies that during the propagation of light at a frequency ω, nonlinear components of the polarization at frequencies 2ω and 3ω arise and harmonics of the original optical field at 2ω and 3ω are generated. However, with increasing order, a rapid decrease in the susceptibility coefficients is observed. As a consequence, the nonlinear polarization $\mathbf{P}_{NL}$ can be approximated by the quadratic term $\varepsilon_0 \chi^{(2)} \mathbf{E}^2$ for moderate energies. Considering the usual expression of the electromagnetic field

$$\mathbf{E} = \frac{1}{2} \mathbf{E}_0 \, \exp\left[j(\omega t - \mathbf{k}.\mathbf{r})\right] + c.c \tag{5}$$

with $\mathbf{E}_0$, the amplitude and k the wave vector and c.c, the complex conjugate of the formula.

the polarization then becomes equal to:

$$\mathbf{P} = \varepsilon_0 \chi^{(1)} \, \mathbf{E} + \varepsilon_0 \chi^{(2)} \, \mathbf{E}^2 \tag{6}$$

$$\mathbf{P} = \frac{1}{2} \left\{ \begin{array}{l} \varepsilon_0 \chi^{(2)} \, \mathbf{E}_0^2 \\ + \varepsilon_0 \chi^{(1)} \, \mathbf{E}_0 \exp\left[j(\omega t - \mathbf{k}.\mathbf{r})\right] \\ + \frac{1}{2} \varepsilon_0 \chi^{(2)} \mathbf{E}_0^2 \exp\left[2j(\omega t - \mathbf{k}.\mathbf{r})\right] \end{array} \right\} \tag{7}$$

Finally, the net polarization is composed of three components: a continuous one corresponding to the phenomenon of optical rectification, a component at frequency ω (optical polarizability) and a second harmonic term (frequency 2ω) corresponding to the phenomenon known as SHG (second harmonic generation). In SHG, a fundamental wave of amplitude E_ω, angular frequency ω (wavelength λ) and wave vector k_ω passing through a crystal generates a second harmonic wave of amplitude $E_{2\omega}$, angular frequency 2ω (wavelength $\lambda/2$) and wave vector $k_{2\omega}$.

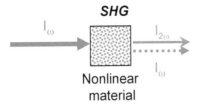

Fig. 9. Second harmonic generation

With distance traveled through the crystal, the second harmonic amplitude varies. The energy generated at frequency 2ω can be obtained by solving the propagation equation:

$$\nabla^2 \mathbf{E} - \mu_0\,\varepsilon_0\,\frac{\partial^2 \mathbf{E}}{\partial t^2} = \mu_0\,\frac{\partial^2 \mathbf{P}}{\partial t^2} \tag{8}$$

with μ_0, the permeability of free space.

The average intensity of an electromagnetic wave of amplitude $\mathbf{E}$ is:

$$I = \frac{\mathbf{E}.\mathbf{E}^*}{2\,\varepsilon_0\,c} \tag{9}$$

with c, the velocity of light. Assuming that the waves are traveling in the z direction and that the conversion efficiency is low (so amplitude of the fundamental is almost unchanged), the second harmonic intensity after a distance ℓ through the crystal is (Armstrong, 1962):

$$I_{2\omega}(\ell) = \frac{\omega^2\,\left(\chi^{(2)}\right)^2\,\ell^2}{2\,\varepsilon_0\,c^3\,n_\omega^2\,n_{2\omega}}\;\frac{\sin^2\left(\dfrac{\Delta k\,\ell}{2}\right)}{\left(\dfrac{\Delta k\,\ell}{2}\right)^2}\;I_\omega^2 \tag{10}$$

with :

n_ω, the refractive index of the crystal at angular frequency ω

$n_{2\omega}$, the refractive index of the crystal at angular frequency 2ω

Δk , the phase mismatch: $\Delta k = k_{2\omega} - 2 k_\omega = \dfrac{2\,\omega}{c}(n_{2\omega} - n_\omega)$;

$I_{2\omega}$ depends on several parameters and among them on the value of Δk . The case of $\Delta k=0$ corresponds to materials called "phase-matchable materials" for which the longer the distance traveled inside the crystal, the greater the SHG intensity (i.e. large particles will generate a better SHG signal). Phase-matchable materials such as potassium diphosphate (KDP) are used to double high power lasers or as a nonlinear standard. But most materials are "non phase-matchable" (ie. $\Delta k \neq 0$) with as a consequence an intensity value oscillating with ℓ . $I_{2\omega}$ will reach a maximum only for the discrete values of ℓ given by:

$$\ell_c = \frac{\pi}{\Delta k} = \frac{\lambda}{4(n_{2\omega} - n_\omega)} \tag{11}$$

ℓ_c is called the "coherence length".

This implies that the SHG intensity will be optimized for particles of size equal to ℓ_c or equal to an odd multiple of ℓ_c. The consequence of the coherence length will be discussed later. $I_{2\omega}$ depends also on $\chi^{(2)}$, the second order nonlinear susceptibility, and this is the main point for the detection of conglomerates. 27 numbers constitute the electro-optic components of $\chi^{(2)}$ which is a 3rd rank tensor. However, the number of independent coefficients can be reduced to 10 if the absorption of the material is negligible at ω and 2ω. In this case, the tensor is invariant by circular permutation of its three indices.

Moreover, the number of independent coefficients non equal to zero can be determined by tacking into account the symmetry elements of the 32 crystallographic classes. In particular, all the components of $\chi^{(2)}$ are null for centrosymmetric structures. Thus, in crystals with a centre of inversion, all the components of $\chi^{(2)}$ tensor are zero. Consequently, these types of crystals can not exhibit a SHG signal. According to the Kleinman symmetry rules (Kleinman, 1962) three chiral point groups (NC) present also a null $\chi^{(2)}$ (422, 622 and 432). However, Kleinman symmetry is not always applicable. These different possibilities are summarized in figure 10.

As a result, the absence of a center of symmetry can be determined via the detection of a SHG signal. Therefore, this test was chosen as a pre-screening method for spotting conglomerates.

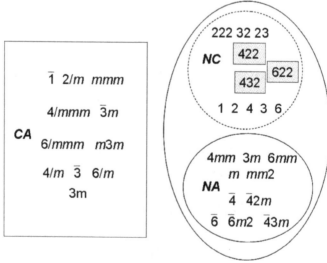

☐ : centrosymmetric point groups (no SHG activity)
0 : non centrosymmetric point groups (SHG activity)
▓ : no SHG activity if Kleinman symmetry is applicable

Fig. 10. SHG activity among the 32 crystallographic classes

4.2.2 Experimental set-up

The experimental set up proposed by Kurtz and Perry (Kurtz et al, 1968) gives quick information about the SHG activity of powder samples. Because the majority of the pharmaceutical crystalline samples are available in the form of powders, this set-up was chosen for the pre-detection of conglomerates.

Figure 11, shows the experimental set-up used for the SHG test. The laser is a Nd:YAG Q-switched laser operating at 1.06 μm. It delivers 360 mJ pulses of 5 ns duration with a repetition rate of 10 Hz. An energy adjustment device is made up of two polarizers (P) and a half-wave plate ($\lambda/2$).

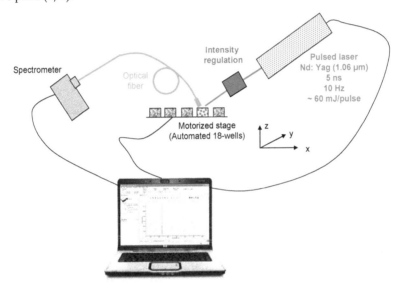

Fig. 11. Experimental Set-up

It allows the incident energy to vary from 0 to circa 200 mJ per pulse. A RG1000 filter is situated after the energy adjustment device to removes light from laser flash lamps. The samples (18-wells plate filled with the various powders) are placed on a motorized stage and irradiated with 60 mJ pulses (beam diameter of 4 mm). The signal generated by the sample (diffused light) is collected into an optical fiber (500 μm of core diameter) and directed onto the entrance slit of a spectrometer. A boxcar integrator allowed an average spectrum (spectral range 250-700 nm with a resolution of ± 0.2 nm) to be recorded over 1 second (10 pulses).

4.2.3 Accuracy of the SHG method

Because of the low occurrence of conglomerates, it would be extremely detrimental to fail to spot one. Conversely, it is not awkward to have a limited number of false positive responses. The diagram presented on figure 12 gathers the various situations that can be encountered during the SHG prescreening. It can be used to evaluate the consistency of the SHG response observed for a given sample.

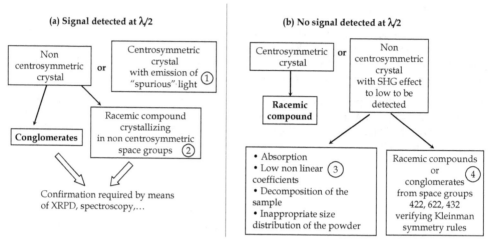

Fig. 12. Decision diagram for SHG test

Panel (a) of the diagram corresponds to the samples for which a signal has been observed. It is necessary to distinguish between two cases: (1) the observation of a "non SHG" signal resulting from optical phenomena such as two-photon fluorescence (TPF) or other photoluminescent processes and (2) the observation of a "real SHG" signal resulting from racemic compounds crystallizing in non-centrosymmetric space groups.

False positive response arising from case 1 can be simply encountered. Indeed, TPF signals present a spectral bandwidth much broader than that of the SHG signal (cf. figure 13).

A base line correction around the SHG wavelength allows the distinction between light resulting from the SHG process itself (signal of interest) and "spurious" light. This simple procedure strongly limits the occurrence of false positive responses.

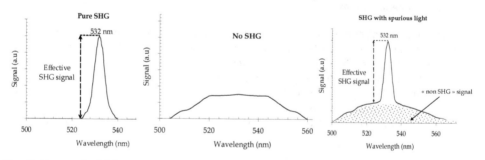

Fig. 13. Removal of false positive responses

As mentioned in table 1, it is possible for a racemic mixture to crystallize in a non-centrosymmetric space group. These types of crystals generate a SHG response and can also be considered as false positive. They constitute case 2. It is also worth mentioning that slightly disordered racemic structures and solid solutions can fall in this category. The influence of the degree of disorder (or pseudo-symmetry) of the structure is difficult to establish so as the consequences on the $\chi^{(2)}$ coefficients. However, the detection of partial solid solutions can be of interest since some of them can be potentially adequate for preferential crystallization (Wermester et al, 2007).

Panel (b) of the diagram concerns samples that do not present any signal at $\lambda/2$. The critical situation of non detected conglomerates is considered in case 3. The main reasons of the "no-detection" are an SHG effect too weak to be detected (due to for example to low non linear coefficients consequence of a low hyperpolarizability of the molecule) or an inappropriate crystal size distribution. The second harmonic signal generated by a crystalline powder is the sum of the contribution of each individual particle (electromagnetic fields are uncorrelated). Because the fundamental beam passes through a large number of particles for which a random orientation is assumed, the intensity of the SHG beam is not easy to optimize and depends on the particle size (especially in non phase matchable materials). To preserve the reliability of the method, too fine particles as those present in submicronic or nanocrystalline powders should be avoided (coherence length is in order of magnitude of several microns for most materials). To limit the number of undetected conglomerates, it is also necessary for the experimental set-up to present a detection level better than 1/100 of the SHG signal generated by the quartz powder (standard non phase-matchable material for SHG measurements – mean diameter 50 μm). This condition is in most cases considered as sufficiently constraining. To be exhaustive, let us mentioned that a non detectable (or a decreasing) signal can also be the consequence of an absorption in the sample at the wavelength of irradiation (λ) or at the wavelength of the re-emitted radiation ($\lambda/2$). This absorption can generate a decomposition of the sample when exposed to the laser beam. The use of an alternative laser source must then be envisaged.

The last case (Case 4) concerns chiral crystals associated to point groups 422, 622 and 432. These should be SHG inactive due to the application of Kleinman permutation rules. However, experience refutes the general applicability of these rules. Recently, SHG activity has been observed in N-acetyl-methylbenzylamine which crystallizes as a conglomerate in space group $P4_12_12$.

Considering all these elements, spotting a SHG active substance can not guaranty the existence of a conglomerate. Only a complementary study of the samples by conventional methods can conclude on the conglomerate nature or not. That is why this method is proposed as a prescreening technique only. However, SHG presents numerous benefits listed below:

- Only a small quantity of the racemic mixture is necessary (ca. 20 mg typically).
- There is at this first stage of pre-screening no need for comparison between results of SHG tests on the racemic mixture and the pure enantiomer. The screening can be undertaken even when the pure enantiomer is not available and thus be carried out at an early stage of the development of the molecule.
- The response is instantaneously delivered; it is thus conceived to be a true high throughput pre-screening method which allows on a short period of time to test numerous derivatives.
- It is a priori a non destructive method.
- It is cheap and can be fully automated: a high throughput device including motorized sample holders and a computer assisted treatment of the spectra should allow in a realistic way to select the "good" candidates with a 50% probability or more.

4.3 Rationalization of conglomerate detection

When a positive SHG signal is obtained, the conglomerate nature of the compound has to be confirmed via XRPD. However, it is worth mentioning that some bias in the conglomerate detection can be introduced at this stage if a strict control of the crystallization conditions is not applied. To avoid misinterpretation of the results, the diagram of figure 14 is proposed for an optimization of conglomerate detection (Gonella, 2011).

Once the SHG prescreening has led to suspect the existence of a conglomerate (step 1 completed), the next step consists in comparing the XRPD patterns of the racemic mixture and the corresponding enantiomer. At this stage it is imperative to apply the same crystallization conditions for the two samples to avoid discrepancies due to polymorphism or solvation / desolvation. It concerns in particular the nature of the counter-ion or cocrystal former, the stoichiometry, the temperature (during the crystallization but also during the SHG detection), the nature of the solvent, etc.

Once both compounds are available (step 2), a precise comparison of the XRPD patterns can be pursued.

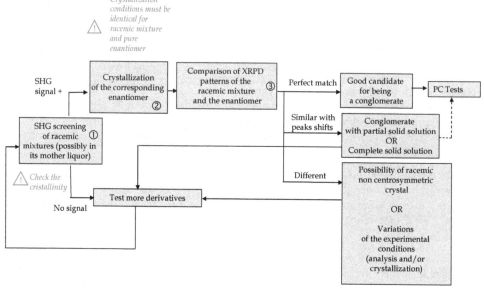

Fig. 14. Workflow for conglomerate detection

Three different cases can be encountered on step 3 : perfect match, similar diffractograms with some peak shifts or different diffractograms). On Figure 15 are presented the XRPD patterns obtained for three different compounds. The upper traces correspond to the racemic mixture (1a, 2a, 3a) and the lower ones to the pure enantiomer (1b, 2b, 3b).

Diffractograms 1a and 1b perfectly match which permits to conclude that compound 1 is with a high level of confidence a conglomerate. PC tests can be undertaken. Diffractograms 2a and 2b present numerous differences. Compound 2 is unlikely to be a conglomerate

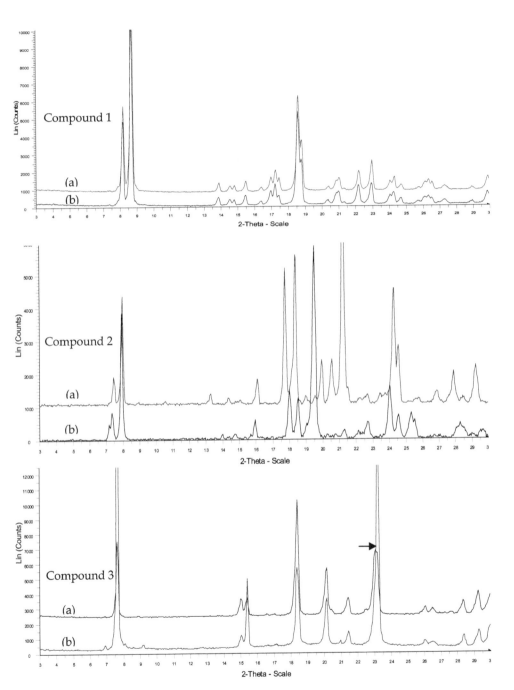

Fig. 15. Comparison of X-Ray patterns for the detection of conglomerate

forming system. However, it is worth mentioning that interesting conglomerates may exist only in equilibrium with their mother liquor (because of one or several of the following reasons: non congruent solubility, solvates with an efflorescent character, solvates which are hydrolyzed by moistures, CO_2 sensitive solid, etc). So it is of primary importance to check if the conditions used during the SHG test are identical to those used for XRPD. Indeed, due to the quickness of the SHG test, an efflorescent solvate could be detected as a conglomerate. Its desolvation prior or concomitantly to the XRPD analysis could then lead to contradictory results. To overcome these problems, SHG and XRPD tests should be run under strictly identical conditions of crystallization.

Diffractograms 3a and 3b are extensively similar but show slight shifts in 2θ positions of some peaks. This couple of derivatives deserves more investigations. Nevertheless, it is likely that this system exhibits at least partial solid solutions.

5. Conclusion

The objective of this chapter was a better understanding of chiral discrimination in the solid state and of the mechanisms involves during recrystallization of enantiomers.

We described two methods of enantiopurification and underlined the benefits of preferential crystallization. Because the detection of a conglomerate is an essential step of this process, we detailed a recently developed technique of prescreening of conglomerates. We finally proposed a workflow to follow in order to optimize the detection of conglomerates and avoid bias that can be induced by different crystallization parameters.

6. References

Armstrong, J. A., Bloembergen, N., Ducuing, J., & Pershan, P. S. (1962). Interactions between light waves in a nonlinear dielectric. *Physical Review*, Vol. 127, No. 6, (September 1962), pp. 1918–1939

Belsky, V. K. & Zorkii, P. M. (1977), Distribution of organic homomolecular crystals by chiral types and structural classes, *Acta Crystallographica Section A*, Vol. 33, No. 6, (November 1977), pp. 1004–1006, ISSN 1600-5724

Brock, C. P., Schweizer, W. B., & Dunitz, J. D. (1991), On the validity of wallach's rule: on the density and stability of racemic crystals compared with their chiral counterparts, *Journal of American Chemical Society*, Vol. 113, No. 26, (December 1991), pp. 9811–9820, ISSN 0002-7863

Claborn, K., Puklin-Faucher, E., Kurimoto, M., Kaminsky, W., & Kahr, B. (2003), Circular dichroism imaging microscopy: application to enantiomorphous twinning in biaxial crystals of 1,8-dihydroxyanthraquinone, *Journal of the American Chemical Society*, Vol. 125, No. 48, (December 2003), pp. 14825–14831, ISSN-0002-7863.

Collet, A. (1999), Separation and purification of enantiomer by crystallization methods, *Enantiomer*, Vol. 4, No. , (1999), ISSN

Coquerel, G., Petit, M.-N., & Bouaziz (1995). Method of resolution of two enantiomers by crystallisation. *PCT Patent WO 95/08522.*

Coquerel, G.; Petit, M.-N.; Bouaziz, R.(1994) Procédé de dédoublement (AS3PC) de deux antipodes optiques par entraînement polythermique programmé et auto-ensemencé. PCT N°94/01.107, 22/09/1994, 1994.

Coquerel, G. (2000), Review on the heterogeneous equilibria between condensed phases in binary systems of enantiomers, *Enantiomer*, Vol. 5, No. 5, pp. 481-498, (Mai 2000), ISNN 1024-2430

Coquerel, G. (2007). Preferential Crystallization, In : *Novel optical resolution technologies - Topics in Current Chemistry*, Sakai, Kenichi et al (Eds), pp. 1–51, Springer, ISBN 978-3-540-46317-7, Berlin.

Dalhus, B. & Görbitz, C. H. (2000), Non-centrosymmetric racemates: space-group frequencies and conformational similarities between crystallographically independent molecules, *Acta Crystallographica Section B*, Vol. 56, No. 4, (August 2000), pp. 715–719, ISSN 1600-5740

Flack, H. (2003), Chiral and achiral crystal structures, Helvetica Chimica Acta, Vol.86, No.4, (January 2003), pp. 905–921, ISSN 1522-2675

Flack, H. D. & Bernardinelli, G. (2003), The mirror of galadriel: looking at chiral and achiral crystal structures, *Crystal Engineering*, Vol. 6, No. 4, (December 2003) pp. 213–223, ISSN 1463-0184

Galland, A., Dupray, V., Berton, B., Morin-Grognet, S., Sanselme, M., Atmani, H., & Coquerel, G. (2009), Spotting conglomerates by second harmonic generation. *Crystal Growth & Design*, Vol. 9, No. 6, (May 2009), pp. 2713–2718, ISSN 1528-7483

Gernez, D. (1866) Séparation des tartrates droits et des tartrates gauches à l'aide de solutions saturées, *Compte-rendus de l'Académie des Sciences*, Vol. 63, (July-December 1866), pp. 843-888

Gonella, S., Mahieux, J., Sanselme, M., & Coquerel, G. (2011). Spotting a conglomerate is just halfway to achieving a preparative resolution by preferential crystallization. *Organic Process Research & Development*, In press, ISSN 1083-6160.

Helmreich, M.; Niesert, C.-P.; Cravo, D.; Coquerel, G.; Levilain, G.; Wacharine-Antar, S.; Cardinaël, P, (2010), Process for enantiomeric separation of racemic dihydro-1, 3, 5 triazines via preferential crystallization. *PCT patent WO2010109015 (A1), 2010-09-30*

Jacques, J., Collet, A., & Wilen, S. (1994). Enantiomers, Racemates and Resolutions (3rd Edition), Kriger Pub. Co., ISBN 0894-648764, Malabar Florida, USA.

Levilain, G., Rougeot, C., Guillen, F., Plaquevent, J-C., Coquerel, G. (2009) Attrition enhanced preferential crystallization combined with racemization leading toredissolution of the antipode nuclei, *Tetrahedron: Asymmetry*, Vol. 20, No. 24, (December 2009), pp. 2769–2771, ISSN 0957-4166

Levilain, G., Coquerel, G. (2010) Pitfalls and rewards of preferential crystallization, *CrystEngComm*, Vol. 12, No. 7, (May 2010), pp. 1983-1992, ISSN 1466-8033

Jungfleish, M. E. (1882) , *Journal of Pharmaceutical Chemistry*, Vol. 5, pp. 346.

Kaptein, B., Elsenberg, H., Grimbergen, R.F.P., Broxterman, Q.B, Hulshof, L.A., Pouwer, K.L. & Vries, T.R. (2000) Dutch resolution of racemic 4-hydroxy- and 4-fluorophenylglycine with mixtures of phenylglycine and (+)-10-camphorsulfonic acid, *Tetrahedron: Asymmetry*, Vol. 11, No. 6, (April 2000), pp. 1343-1351, ISSN 0957-4166

Kleinman, D. A. (1962). Theory of second harmonic generation of light. *Physical Review*, Vol. 128, No. 4, (1962) pp. 1761–1775.

Kostyanovsky, R. G., Kostyanovsky, V. R., & Kadorkina, G. K. (2009) The enigma of a (±)-tartaric acid-urea cocrystal, Mendeleev Communications, Vol. 19, No.1, (2009), pp. 17- 18, ISSN 0959-9436

Kuroda, R. & Honma, T. (2000) Cd spectra of solid-state samples, Chirality, Vol. 12, No. 4, (April 2000), pp. 269–277, ISSN 0899-0042

Kurtz, S. & Perry, T. (1968) A powder technique for the evaluation of nonlinear optical materials, Journal of Applied physics, Vol.39, No. 8, (July 1968), pp.3798–3813, ISSN 0021-8979.

Lorenz, H., Perlberg, A., Sapoundjiev, D., Elsner, M. P., & Seidel-Morgenstern, A. (2006), Crystallization of enantiomers, Chemical Engineering and Processing: Process Intensification, Vol. 45, No. 10, (April 2006), pp. 863 – 873, ISSN 0255-2701

Marchand, P., Lefebvre, L., Querniard, F., Cardinael, P., Perez, G., Counioux, J-J & Coquerel, G. (2004), Diastereomeric resolution rationalized by phase diagrams under the actual conditions of the experimental process, Tetrahedron: Asymmetry, Vol.15, No.16, (August 2004), pp. 2455-2465, ISSN 0957-4166

Marckwald, W., (1896) Ueber ein bequemes Verfahren zur Gewinnung der Linksweinsäure, Berichte der deutschen chemischen Gesellschaft, Vol.29, No.1, (April 1896), pp.42–43

Ndzié, E., Cardinael, P., Schoofs, A. R., & Coquerel, G. (1997), An efficient access to the enantiomers of [alpha]-methyl-4-carboxyphenylglycine via a hydantoin route using a practical variant of preferential crystallization as3pc (auto seeded programmed polythermic preferential crystallization), Tetrahedron: Asymmetry, Vol. 8, No. 17, (September 1997), pp. 2913–2920, ISSN 0957-4166

Kocher-Becker, U. , Kocher, W. & Ockenfels, H. (1982), Teratogenic activity of a hydrolytic thalidomide metabolite in mice, Naturwissenschaften, Vol. 69, No. 4, (Avril 1982), pp. 191-192, ISSN 0028-1042

Pasteur, L. (1853), Transformation des acides tartriques en acide racémique - Découverte de l'acide tartrique inactif. Nouvelle méthode de séparation de l'acide racémique en acides tartriques droit et gauche. Compte-rendus de l'Académie des Sciences, Vol. 37, (Juillet 1853), pp.162-166

Pope, W.J. & Peachey, S.J. (1899), The application of powerful optically active acids to the resolution of externally compensated basic substances. Resolution of tetrahydroquinaldine, Journal of Chemical Society Transactions, Vol. 75, (January 1899), pp. 1066-1093, ISNN 0368-1645

Renou, L., Morelli, T., Coste, S., Petit, M.-N., Berton, B., Malandain, J.-J. & Coquerel, G., (2007), Chiral discrimination at the solid state of the methyl 2-(diphenylmethylsulfinil)acetate, Crystal Growth & Design, Vol. 7, No. 9, (August 2007), pp.1599-1607, ISSN 1528-7483

Stinson, S.C. (2000). , Chiral Drugs, Chemical and Engeniering news, Vol. 78, No. 43, pp. 55-78 (October 2000), ISSN 0009-2347

Subramanian, G., (2006). Chiral Separation Techniques: A Practical Approach (3rd – 18th October), Wiley-VCH Verlag GmbH, ISBN: 978-3-527-31509-3, Weinheim, Germany.

Vries, T., Wynberg, H.,Van Echten, E.A, Koek, J.A, Ten Hoeve, W.A., Kellogg, R.M., Broxterman, Q.B., Minnaard, A.B, Kaptein, B.B, Van Der Sluis, S., Hulshof, L. & Kooistra, J. (1998), The family approach to the resolution of racemates, Angewandte Chemie - International Edition, Vol. 37, No 17, (September 1998), pp. 2349-2354, ISSN 1433-7851

Wermester, N., Aubin, E., Pauchet, M., Coste, S., & Coquerel, G. (2007). Preferential crystallization in an unusual case of conglomerate with partial solid solutions. Tetrahedron: Asymmetry, Vol. 18, No. 7, (2007), pp. 821–831, ISSN 0957-4166

10

Crystal Forms of Anti-HIV Drugs: Role of Recrystallization

Renu Chadha, Poonam Arora, Anupam Saini and Swati Bhandari
University Institute of Pharmaceutical Sciences, Panjab University, Chandigarh,
India

1. Introduction

Understanding and controlling the solid-state chemistry of active pharmaceutical ingredients (APIs) is an important aspect of drug development process. APIs can exist in a variety of distinct solid forms, including polymorphs, solvates, hydrates, salts, cocrystals and amorphous solids. Most APIs are purified and isolated by crystallization from an appropriate solvent during the final step in the synthetic process. A large number of factors can influence the crystal nucleation and growth during this process, including the composition of the crystallization medium and the processes used to generate supersaturation and promote crystallization. For development of a pharmaceutical product, it is generally accepted that the stable form should be identified and chosen for development. However, the stable crystal form of the parent compound may exhibit inadequate solubility or dissolution rate resulting in poor oral absorption, particularly for water insoluble compounds whereas a metastable form might have advantageous properties. Moreover, the metastable polymorphs constitute local minima in the energy landscape (Figure 1) with the thermodynamically stable form being the absolute minimum at a given temperature and pressure. Thus, the search for absolute minimum and energy differences between the local minima of the drug substances is the goal of material and formulation scientists in the pharmaceutical industry. While significant efforts are made by drug development groups to identify and characterize thermodynamically stable crystal forms early in development, there are many instances where new crystal forms have been discovered later in development process. The late emergence of thermodynamically stable crystal form is often explained by Ostwald's law of stages which states that the least stable crystal form is likely to crystallize first. Metastable forms appear first during crystallization process as their crystallization kinetics is faster than those of stable forms but eventually transform into a stable form. It is important to study the transformations, because the sudden appearance or disappearance of a polymorphic form in pharmaceutical products can lead to serious consequences.

Therefore, it is of utmost importance to control the crystal formation and produce a desired form. Discovery and characterization of the diversity of solid forms of a drug substance provide options from which to select a form that exhibits the appropriate balance of the critical properties for development into the drug product. Lately, the crystal engineering approaches utilizing high throughput techniques have been applied to crystalline materials to fruitfully generate various crystal forms of pharmaceutical compounds. The ability to engineer pharmaceutical materials with suitable solubility characteristics whilst maintaining suitable physical and chemical stability provides a driving force to utilize modern analytical tools to generate new crystal modifications.

Drugs with multicomponent crystalline phases such as cocrystals also carry desired drug properties similar to single-component polymorphs. The pharmaceutical cocrystals is defined as multicomponent molecular complex where one of the components is an investigational or marketed drug molecule (the active pharmaceutical ingredient or API) and the second component (the coformer) is a safe chemical for human consumption selected from the GRAS list of the US FDA (generally regarded as safe additive chemicals by the Food and Drugs Administration). The two components are present in a definite stoichiometric ratio and interact through noncovalent interactions, predominantly hydrogen bonds. Cocrystals have been found to offer an attractive platform to improve the solubility and dissolution rate of pharmaceuticals without compromising on the stability of the solid form. Many pharmaceutical companies are working actively on cocrystals and this is reflected by the growing number of publications and patent applications for co-crystals in recent years.

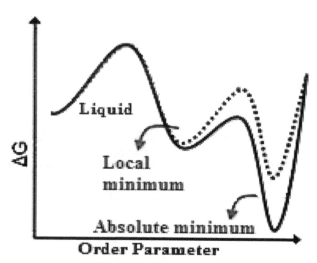

Fig. 1. Energy landscape for polymorphic forms.

The identification and characterization of diverse crystalline forms of the drug substances has become imperative for the solid state chemists in order to select the appropriate form to ensure that the product performance with respect to manufacturability, stability and bioavailability remains unchanged. Therefore, this chapter deals with the effect of crystalline state of pharmaceutical material on its physicochemical properties demonstrated through several case studies describing the phenomenon of polymorphism and a particular attention will also be paid to cocrystallization which is emerging as an important technique to generate crystal forms with improved physicochemical properties (Scheme 1).

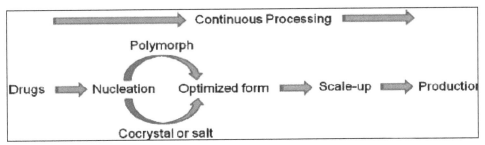

Scheme 1. Scheme representing the path followed by drugs from nucleation to production

1.1 Effect of crystal forms on physicochemical properties

It is well established that different solid forms of drug molecules exist and can affect pharmaceutical drug products with respect to physicochemical properties. It is the variation in the physical and chemical properties of drug molecules that makes polymorphism such a potentially important issue for the pharmaceutical industry. The APIs with abundant hydrogen bond sites and molecular flexibility may be manipulated by proper choice of solvent to form a specific crystal form that has different arrangements or conformations of the molecule in the crystal lattice. Changing the arrangement of the molecules in the crystal lattice changes the solid state properties affecting its solubility, stability, dissolution rate and bioavailability. Different crystal forms can have different rates of uptake in the body, leading to lower or higher biological activity than desired. As a result, significant effort is placed in identifying suitable solid forms of drug substances for use in pharmaceutical drug products.

1.2 Crystallization process

The crystallization process of polymorphous crystals is concerned with evolution of crystalline state from solution or melts and is composed of competitive nucleation and growth. However, crystallization is a complex process which starts with the appearance of most soluble form and ends with transformation to stable form. Alternatively, the nucleation of the stable form can be initiated by the dissolution of this metastable form and growth of the stable form continues until the solubility of stable forms is reached. In many instances the metastable form with desirable properties may precipitate out which is stable for years. To selectively crystallize polymorphs, the mechanism of each elementary step in crystallization process needs to be clear in relation to operational conditions and the key controlling factors. The nucleation process is the most important for control of polymorphous crystallization. Thus control over solid form throughout the drug development process is of paramount importance.

A number of factors (Figure 2) affect polymorphic behavior of a pharmaceutical solid and type of solvent has a major factor in polymorphic selectivity and crystal morphology.

1.3 Effect of solvent on crystal form

Polymorph selectivity is primarily based on the polarity of the solvent. Thus a systematic approach for selecting the right solvent is beneficial for better experimental design in control of crystallization. This effect arises from the solvent–solute interaction at the molecular level and have been explained by few of the researchers. The solvent–solute interactions during cluster formation for nucleation and growth significantly affect the ultimate crystal structure and morphology. If the solvent–solute are strongly bonded at a special crystal face, the rate-limiting step of growth would be the removal of the solvent from that face. In this case, the bonded surface grows slowly or does not grow.

However, Threlfall showed that if crystallization occurs in a region that is supersaturated with respect to one polymorph (the less soluble form) and under-saturated with respect to the other one, the solvent has no influence on the nucleation of the polymorphs and the thermodynamics will lead the process toward the production of the less soluble polymorph (Threlfall T., 2000). Moreover, the thermodynamically stable polymorph is the most stable form irrespective of the type of the solvent used. Since the thermodynamic stability of different polymorphs does not change with type of the solvent, then the solvent effect on polymorphism is attributed to the kinetic parameters.

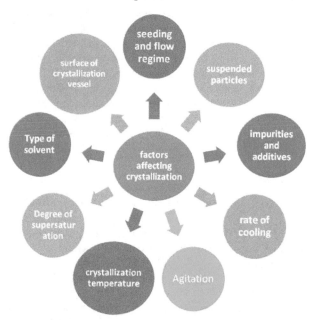

Fig. 2. Factors which affect the crystallization of a drug substance

1.4 Other factors

Within the same solvent system, many factors are known to influence the crystal habit including supersaturation, cooling rate and agitation. At a constant temperature, supersaturation has a direct effect on the nucleation rate. As the supersaturation is increased, the rate of nuclei formation is greater than crystal growth and growth occurs mainly in one direction, producing elongated crystals. On the other hand, when there is lesser degree of supersaturation, solute solvent interactions are insignificant, producing platy crystals.

Similarly, rate of cooling alters the crystal habit by its influence on degree of supersaturation. Crystallization at a slower cooling rate produces more symmetric crystals compared with faster cooling. During faster cooling, nucleation is faster than crystal growth rate; therefore, many small crystals appear instead of few crystals growing to sufficiently larger size.

Agitation has also an important effect on the process of crystallization. The aspect ratio (ratio of horizontal maximum and vertical maximum distance of particle) is highest for unstirred conditions than during stirring. The crystals obtained under stirring conditions are fine since stirring facilitates the rate of nucleation by an even distribution of the solute molecules in the solvent. Increased nucleation rate is the result of collision of initial crystals with the stirrer and formation of smaller seeds for further crystallization. Additionally, stirring can also break larger crystals to smaller ones. Thus external appearance of a crystal can be altered by changing the growth environment to suit the requirements.

2. Polymorphism in anti-HIV drugs

This chapter throws light on the different crystalline forms reported so far for nevirapine, efavirenz, lamivudine, stavudine and zidovudine. Preparation and isolation methods, structural characterization and properties of polymorphic/solvatomorphic/cocrystal systems as well as phase transformations are illustrated.

2.1 Ritonavir

A number of studies have successfully demonstrated the appearance of different crystalline forms of some anti-HIV drugs upon recrystallization. An early example being that of ritonavir, marketed as Norvir. The late emergence of a thermodynamically more stable form (Form II) which unexpectedly precipitated from the semisolid capsule formulation led to the removal of the product from the market. The new crystal form (form II) appeared after conversion of metastable crystalline form I (Chemburkar et al., 2000; Bauer et al., 2001; Desikan et al., 2005; Miller et al., 2005). Ritonavir polymorphism was investigated using solid state spectroscopy and microscopy techniques including solid state NMR, near infrared spectroscopy, powder X-ray diffraction and single crystal X-ray analysis. Ritonavir was found to exhibit conformational polymorphism with two unique crystal lattices having significantly different solubility properties. An unusual conformation was found for form II that results in a strong hydrogen bonding network. Although the polymorph (form II) corresponding to the "cis" conformation has a more stable packing arrangement, however, nucleation, even in the presence of form II seeds, is energetically unfavored except in highly supersaturated solutions. The coincidence of a highly supersaturated solution and a probable heterogeneous nucleation resulted in the sudden appearance of the more stable form II polymorph.

Form	Melting point, °C	ΔHfus, J/g	Solid state structure
I*	122	78.2	Monoclinic
II*	122	87.8	Orthorhombic
III	78–82	60.3	Monoclinic
IV	116	59.8	Not assigned
V	97	32.0	Monoclinic

*Bauer *et al.*

Table 1. Comparison of physical parameters of ritonavir crystal forms

This polymorphic shift in ritonavir illustrated the need for early and comprehensive identification of solid-form diversity of this API. The polymorphic behavior of ritonavir was explored by Morissette et al (Morissette et al., 2003) in 2003 using CrystalMax, a high-throughput crystallization platform, with the aim of finding known and novel crystal forms of the drug molecule. Three additional crystalline forms of ritonavir were discovered when about 2,000 screening experiments were carried out (Table 1). These forms were found along with both known forms I and II, which were obtained from previously unreported solvent mixtures. Form III is a crystalline formamide solvate that converts to form V, a previously unknown hydrated phase, upon exposure to aqueous medium. Form V which is a trihydrate obtained from exposing the form III to aqueous conditions, in turn converts spontaneously to needle-like form I crystals. The process of preparing form I from III is an unusual route to a "disappearing polymorph" and provides a novel strategy for control of particle size and morphology. Form IV is a true, unsolvated, metastable previously unreported polymorph of ritonavir. Optical imaging (Figure 3) and *in situ* Raman spectroscopy were used to characterize newly formed crystals. Each of the novel forms found by means of high-throughput crystallization was scaled up to multiple milligram and gram levels. Thus the high-throughput crystallization for solid-form discovery and exploration of large numbers of parallel crystallization trials led to identification of more polymorphic forms of ritonavir.

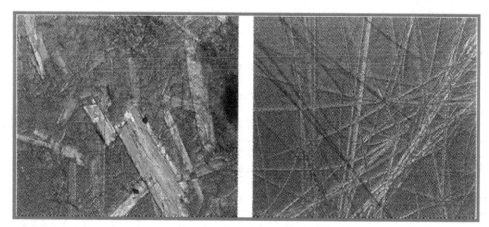

Fig. 3. Video micrograph of crystal Form I (left) and Form II (right).

2.2 Nevirapine

Three polymorphs and many solvatomorphs have been reported for nevirapine, a non-nucleoside reverse transcriptase inhibitor, depending upon the recrystallization method. Specifically, from a crystal engineering viewpoint, the presence of the amide function CO-NH in the nevirapine molecule indicates the possibility of alternative modes of self association, namely via a dimer or a catemer synthon, leading to crystal polymorphism, whereas interaction with the solvent molecules having complementary donor and acceptor functions result in formation of solvates, significantly extending the solid state chemistry of the drug. Form I was prepared by recrystallizing nevirapine from alcohols, ethers, esters or their mixtures while refluxing from toluene, n-butanol or methyl-isobutyl ketone and subsequent cooling to 0-10 °C yielded form II. Form III was prepared by refluxing nevirapine in chloroform and using dichloromethane as antisolvent to the reaction mixture (Reguri and Chakka, 2005, 2006). However, many other experiments to investigate the existence of any other thermodynamically more stable form at room temperature, led to appearance of different solvatomorphs of nevirapine with varying stoichiometries depending upon the solvents selected. The first report on preparation of different solvatomorph by Pereira et al appeared in 2007 (Pereira et al., 2007). Six different solvates of nevirapine with different morphology were obtained by saturating the solvent systems with drug at room temperature and cooling in refrigerator. Despite different morphologies the DSC profiles did not show any relevant differences for raw material and other forms at crystal fusion peaks. However, a thermal event was observed below the melting temperature in NEV 3, NEV4 and NEV6 indicating the loss of some solvent molecules. The presence of solvent is confirmed by TGA and Karl fischer analysis. The authors have reported the crystal structure of two solvate forms of nevirapine, hemihydrate (NEV3) and hemiethylacetate (NEV4) (Figure 4).

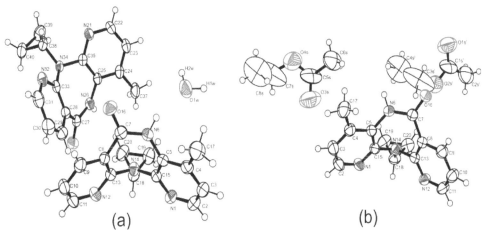

Fig. 4. Ortep representation of structure of (a) nevirapine hemihydrate and (b) nevirapine hemiethyl acetate solvate.

Caira et al have also reported five different solvates of nevirapine (Caira et al., 2008). The initial characterization of all the forms has been performed utilizing DSC and TGA. The results show that the drug molecule displays significant variability in its modes of self assembly while accommodating the different solvent molecules. The DSC results show less thermal stability for toluene solvate while ethylacetate solvate, dichloromethane solvate, hydrate and 1,4-dioxane solvate were stable. The ethylacetate solvate and dichloromethane solvate are isostructural and their dimmers are packed in identical fashion, generating continuous channels parallel to the b-axis that accommodate the solvent molecules (Figure 5). However, in the toluene solvate containing a larger guest molecule, the host dimers are packed in a different mode, but the guest molecules are again situated in channels.

The effect of series of alcohols on solvate formation capability of nevirapine molecule has also been illustrated by Caira et al (Caira et al., 2010). The structures of all the solvates were based on a common isostructural framework comprising centrosymmetric hydrogen-bonded nevirapine dimers and contain a common channel parallel to the crystal b-axis in the series which accommodates the various solvent molecules. Thermogravimetric results yielded a guest–host ratio close to 0.5 for the 1-butanol solvate and a steady decrease in this ratio from 0.43 to 0.32 for other solvates. This anomalous stoichiometric variation was resolved following successful X-ray analysis of 1-butanol solvate which revealed that the length of disordered 1-butanol molecule is proportionate with the channel cavity, resulting in a stoichiometric association while in other solvates significant disorder for the solvent molecules was observed which is attributed to their increasing chain lengths being disproportionate with the channel cavity.

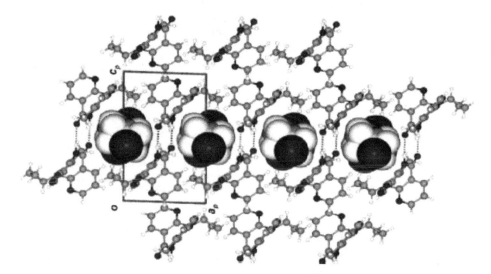

Fig. 5. Inclusion of ethyl acetate molecules within channels in crystals of solvate of nevirapine.

In the view of tendency of solvate formation of nevirapine molecule, other workers have also prepared and characterized the various solvates of nevirapine. In one of the studies, the authors have prepared the solvates by dissolving excess drug in selected solvent systems at 60 °C. However, the choice of solvent system was based on the polarity index of the solvents (Sarkar et al., 2008). Five different crystal forms have been obtained along with an amorphous form. The crystallization of nevirapine under a variety of crystallization conditions resulted in a change in the crystal habit of the drug without change in the internal crystal lattice as suggested by their similar enthalpy of fusion. The mass loss from the TGA was found to be negligible for all the solvates in comparison to the theoretical mass loss indicating that solvents used for crystallization formed weak solvates in this study. However, chadha et al have determined the binding energy of the solvent in the crystal lattice using differential scanning calorimetry which is found to be higher than the enthalpy of vapourization of the corresponding solvent for all the solvates except for toluene solvate indicating that the solvent molecules are tightly bound into the crystal lattice of nevirapine molecule (except in case of toluene solvate) (Chadha et al., 2010). These authors have also calculated the enthalpy of solvation by determining enthalpy of solution of solvate and drug in the solvent which is entrapped in the crystal lattice of the solvate using solution calorimetry technique. The enthalpy of solution when determined in buffer system indicated that out of the six solvates formed ethanol solvate exhibited the maximum ease for molecular release of the solvent molecule from the lattice.

2.3 Efavirenz

The abundant hydrogen bonding sites in efavirenz, another non-nucleoside reverse transcriptase inhibitor, make it a potential candidate to exhibit crystal modifications upon recrystallization. Driven by this aspect, various authors investigated the solid-state structures of recrystallized products of efavirenz. The varying recystallization conditions such as rate of stirring and cooling, antisolvent addition, refluxing/heating, drying under vacuum, and presence of impurities, along with solvents and/or their mixtures with varying polarity have yielded different forms of efavirenz. The patent literature till date reveals 23 different polymorphic forms of efavirenz, one monohydrate and an amorphous form although there is some ambiguity about the actual number of solid forms (polymorphs and solvates) of this API (Radesca et al., 1999, 2004; Sharma et al., 2006; Khanduri et al., 2006; Reddy et al., 2006; Dova, 2008). In these patents, inventors have claimed novel solid forms of efavirenz based on XRPD and DSC analysis. The patent data shows that Form I is the most stable form and all the polymorphs revert to this form under some condition or the other. However, the characterization of these reported forms is not adequate enough to prove them novel.

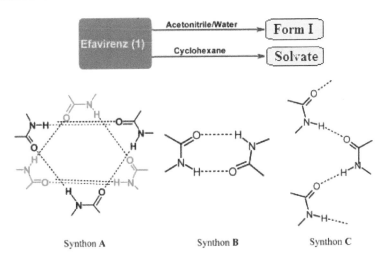

Fig. 6. Solid forms of efavirenz, their transformations and some synthons present in the polymorphs of efavirenz

Recently, two new polymorphc forms structurally characterized by single crystal X-ray diffraction have been reported by Cuffini et al and Ravikumar et al (Cuffini et al., 2009; Ravikumar et al., 2009). These two forms do not correspond to the Form I reported in various patent.

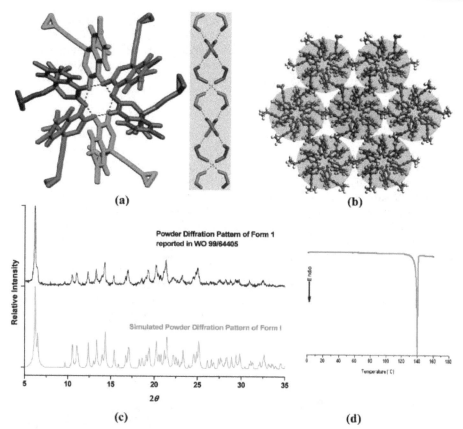

Fig. 7. Efavirenz, form I: (a) A view of the double helical chain viewed down the c-axis with the three symmetry independent molecules represented in red, green, and blue. The inset shows the formation of the double helical chains. (b) Hexagonal close packing of helices. (c) PXRD pattern for form 18g and form I. (d) DSC trace for form I.

The structural information of the stable Form I was first reported by Mahapatra et al (Mahapatra et al., 2010). The authors obtained this form (Form I) at the interface of an acetonitrile-water solvent system as well as from methylcyanide-water mixture (Figure 6). They also reported a solvate of efavirenz which was obtained by dissolving the drug in cyclohexane at 60 °C with stirring and leaving the solution undisturbed for 2 days after filtration to get rectangular crystals. The DSC analysis of these two forms showed that the solvate converts to Form I after desolvation. Both the forms have been further characterized using single crystal X-ray analysis. The efavirenz molecule exhibits a noticeable degree of conformational disorder in the cyclopropyl group. After exhaustive study it was found that Form I crystallized in double stranded helices stabilized by N-H··O molecules (Figure 7). However, in the solvate, two molecules of efavirenz form the amide dimer which are stabilized by C-H··π interactions to form a bilayer stacked in three dimensions to form columns. The solvent molecules are lined in these columns and are stabilized by interactions with CF_3 and cyclopropyl residues (Figure 8).

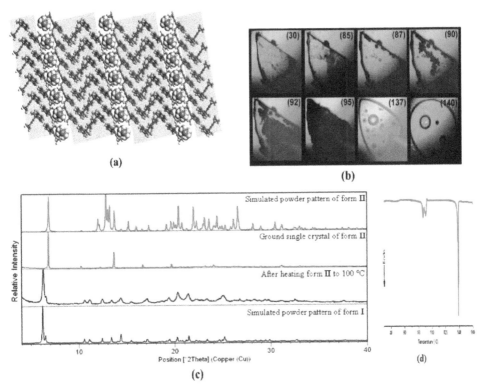

Fig. 8. (a) Crystal packing of efavirenz with the guest cyclohexane molecules represented in the space filling mode (b) HSM images of II (corresponding temperatures are given in parentheses)(c) PXRD patterns of forms I and II and the heated sample of form II (d) DSC plot for form II.

2.4 Lamivudine

Further studies on these categories of drugs have shown the existence of different crystal forms of lamivudine which is a nucleoside analog reverse transcriptase inhibitor. The solid state chemistry of this drug is of significant pharmaceutical interest as the drug is reported to exist in three crystalline forms. The two forms (Form I and II) reported in 1996 were again studied by Harris et al in 1997 (Jozwiakowski et al., 1996; Harris et al., 1997). Later in 2007, a new patent showing the existence of another polymorphic form III appeared (Singh et al., 2007). Michael et al have shown that Form I of lamivudine has been prepared by dissolving Form II in hot water and then adding an equal volume of methanol to reduce the solubility of lamivudine. The Form II has been obtained as a result of a synthetic process. The DSC studies showed heat mediated transformation of Form I to Form II. Besides this, these authors have also calculated the enthalpy of solution from solubility data and compared the results with experimently determined enthalpy of solution by solution calorimetry technique. The enthalpy data revealed that the enthalpy of solution value agrees more closely in systems of low solubility than in system with high solubility.

Harris et al have recrystallized form I of lamivudine as needles from solutions in water, methanol or aqueous alcohols while form II as tetragonal bipyramids on slow recrystallization from dry ethanol and propanol or mixtures of ethanol and less polar organic solvents (Figure 9). Probably, the difference in the polarity of organic solvents led to two different crystalline forms. The authors have used cross-polarization magic angle spinning (CPMAS) NMR to differentiate the two forms. The Form II showed a simple spectrum indicating one molecule in the crystallographic asymmetric unit whereas the spectrum of Form I was found to be extremely complex due to differences in intermolecular packing environment or intramolecular geometry/conformation differences. These results were further confirmed by single crystal X-ray analysis. The Form II was bipyramidal crystals with one molecule in the asymmetric unit and Form I showed five molecules in the asymmetric unit of crystal lattice.

The crystals of Form III are obtained by subjecting the hot saturated solution of lamivudine in water to controlled cooling. The DSC and TGA showed this form to be different from Form I and II. The single crystal X-ray diffraction reveals it to be a hemihydrates with two molecules of water associated with four molecules of lamivudine in a crystal lattice.

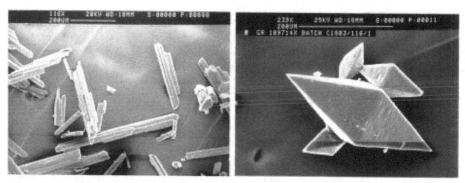

Fig. 9. Scanning electron micrographs of the two forms of Lamivudine

2.5 Stavudine

Stavudine, a thymidine nucleoside reverse transcriptase inhibitor, has been reported to exist in two anhydrous polymorphic forms and one hydrate. Harte et al and Guruskaya et al have reported the single crystal of Form I and Form II respectively (Harte et al., 1991 as sited in Gandhi et al., 2000; Guruskaya et al., 1991 as sited in Gandhi et al., 2000). However, the production of pure Form I i.,s reported by Gandhi et al who have established the conditions of recrystallization governing the formation of this thermodynamically most stable form (Gandhi et al., 2000). The stavudine obtained during the synthesis process is recrystallized from hot organic solvents as the final step in the synthesis. The cooling of hot isopropanol solution from 80 to 70 °C for over an hour and then to 0-5 °C over 1.5 h yielded a mixture of Form I and II. However, Form I was found to be thermodynamically more stable. Therefore, some of the crystallization parameters such as rate of cooling or stirring were studied during recrystallization from isopropanol to selectively obtain Form I. After a lot of experimentation, it was found that slow cooling of hot isopropanol solution reproducibly yielded Form I. Kinetically both forms may be present initially but Form II redissolves and precipitates as Form I with slow cooling.

2.6 Cocrystals

Recently, cocrystallization has emerged as an attractive technique to recrystallize molecular solids that contain two or more distinct chemical components held together by non-covalent interactions. Anti-HIV agents have also been explored by this approach. The cocrystals of efavirenz with 4,4'-bipyridyl and 1,4-cyclohexanedione prepared by recrystallization of their grounded mixture from THF and from a mixture of n-heptane and THF respectively have been reported by Mahapatra et al (Mahapatra et al., 2010). Similarly, the multiple hydrogen bond donor and acceptor groups of lamivudine and zidovudine have been utilized by Bhatt et al for designing their cocrystals (Bhatt et al., 2009). These authors have designed zidovudine cocrystal using retrosynthetic approach where two drug molecules and one molecule of 2,4,6-triaminopyrimidine are held together by a three point synthon which forms basic part of their cocrystal structure (Figure 10).

Besides this, a cocrystal hydrate of lamivudine with zidovudine has been reported. Lamivudine and zidovudine molecules are expected to form synthon II with each other as shown in figure 11. However, during cocrystallization a hydrated 1:1 cocrystal is formed and the synthon formed is the extended IIA synthon rather than II. The observed synthon (IIA) is formed when a molecule of water intervenes in the hydrogen bond pattern of the synthon II. The cocrystallization without use of water resulted in no cocrystal formation in this case, perhaps due to the large repulsions between the carbonyl groups in the two API fragments (Figure 12).

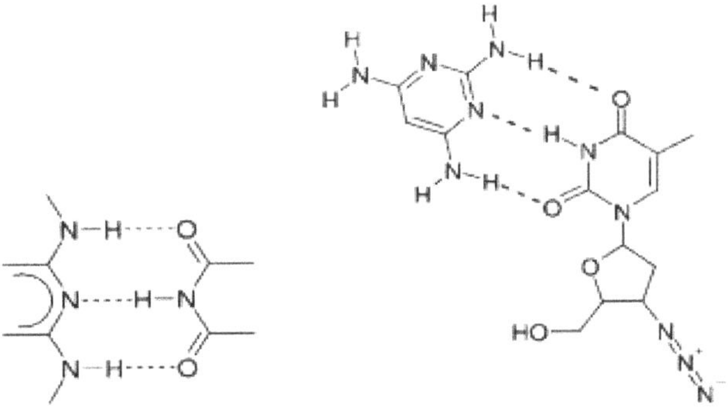

I

Fig. 10. Three-point synthon I and possibility of this synthon between zidovudine and 2,4,6-triaminopyrimidine

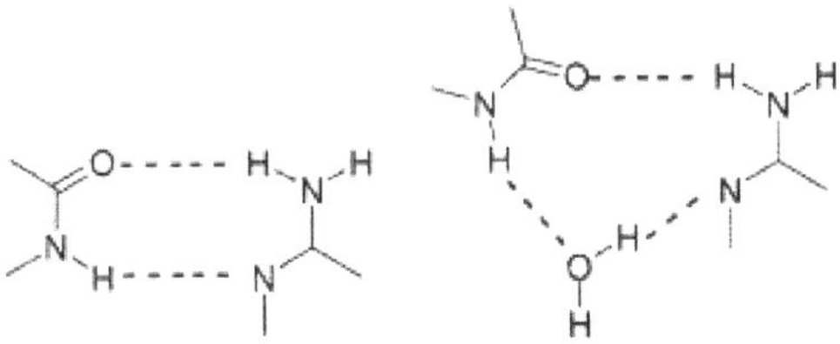

II   IIA

Fig. 11. Predicted two-point synthon II between lamivudine and zidovudine and observed synthon IIa in the hydrated co-crystal

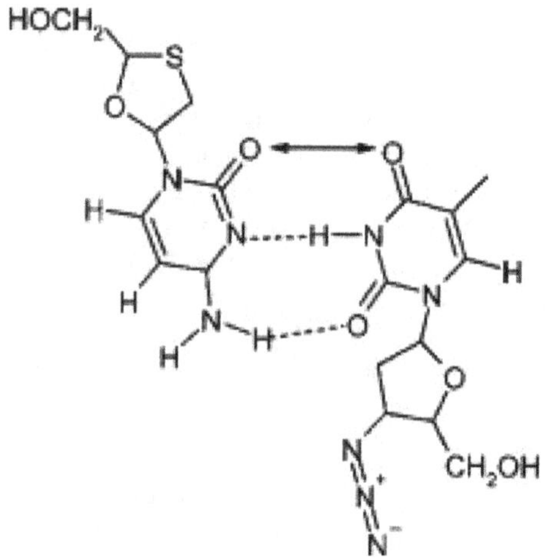

HOCH₂

Fig. 12. Possible carbonyl - carbonyl repulsion in the putative lamivudine-zidovudine co-crystal with non-hydrated synthon II

Another cocrystal of lamivudine is designed and prepared with 3,5-dinitrosalicylic acid based upon synthon containing carboxylic acids and 2-aminopyridines. This cocrystal is an example of acid base interaction with a very complex hydrogen bond pattern. These authors also reported cocrystal of lamivudine with 4-quinolinone in a stoichiometry of 1:1. This cocrystal is stabilized by multiple N-H $\cdots$ O and O-H $\cdots$ O hydrogen bonds. This cocrystal is obtained during the screening process and is not based on any synthon theory. Thus the lamivudine-4-quinolinone cocrystal emphasizes the importance of cocrystal screening to obtain new cocrystals rather than completely depending on the synthon theory.

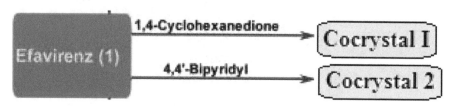

Fig. 13. Cocrystals of efavirenz and their preparation

The possible formation of cocrystals of efavirenz with 1,4-cyclohexanedione and 4,4'-bipyridyl was indicated by X-ray diffraction analysis (Mahapatra et al., 2010). The solids obtained from solvent drop grinding experiment, were subjected to recrystallization in particular solvent systems (Figure 13). Crystals of efavirenz-1,4-cyclohexanedione were obtained from heptane-THF solution. These crystals exhibited two symmetry independent dione molecules in a skewed conformation. One of these two dione molecules makes bifurcated N-H $\cdots$ O/C-H $\cdots$ O hydrogen bonds with two symmetry related molecules of efavirenz, forming a three molecular entity, while the other dione molecule makes linear chains with bifurcated C-H...O hydrogen bonds. These two patterns stacked over one another to form grill-ribbon structure of the cocrystal (Figure 14).

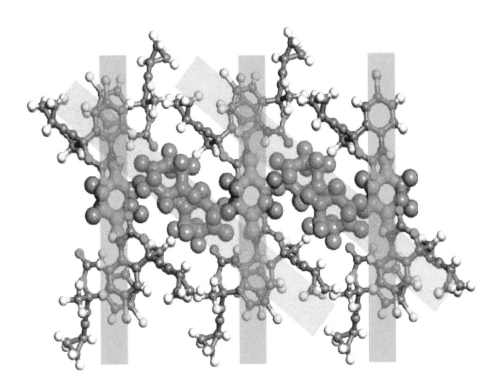

Fig. 14. Cocrystal of efavirenz and 1,4-cyclohexanedione: crystal packing along with the conformation in the native dione crystal (green).

The cocrystal of efavirenz with 4,4′-bipyridyl was obtained by recrystallization from acetonitrile. Unlike the cocrystal with yclohexanedione in which the major synthon couldnot be anticipated, in this cocrystal two distinct heterosynthons were observed. While one of the efavirenz molecules interacted with bipyridyl through cyclic N-H⋯N/C-H·O hydrogen bonds, the other end of pyridine compound made a single point N-H⋯N hydrogen bond with other molecule of efavirenz resulting in pillered assemblies separated by hydrophobic cyclopropyl residues (Figure 15).

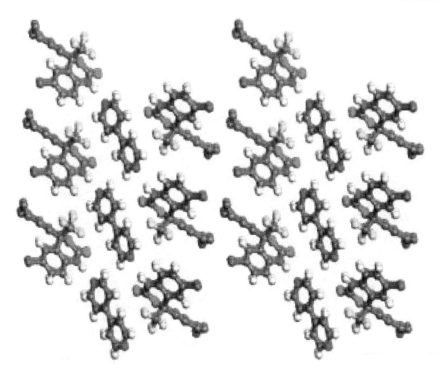

Fig. 15. Crystal packing.of efavirenz- 4,4′-bipyridyl cocrystal

From the above examples it is clear that solvent recrystallization method has been the most common technique to prepare different crystal forms of a compound. However, with increasing demands to accelerate the process of crystal form identification, newer and advanced technologies need to be used as is exemplified by the isolation of three more polymorphic forms of ritonavir by high-throughput crystallization. In addition, cocrystallization technique has also been employing solution based crystallization methods to generate cocrystals of various pharmaceutical compounds. The studies on anti-HIV agents cited here are thus not simply academic exercises but have practical implications in preparation and identification of diverse crystalline forms of pharmaceutical material and can be of considerable benefit in formulation optimization.

3. References

Bauer J., Spanton S., Henry R., Quick J., Dziki W., Porter W., Morris J. Ritonavir: An extraoridinary example of conformational polymorphism. *Pharm. Res.*, 18, 859-866 (2001).

Bhatt, P.M., Azim, Y., Thakur, T.S., Desiraju, G.R. (2009). Co-crystals of the anti-HIV drugs lamivudine and zidovudine. *Cryst. Growth Des.*, 9, 951–957.

Caira, M.R., Stieger, N., Liebenberg, W., De Villiers, M.M., Samsodien H. (2008). Solvent inclusion by the anti-HIV drug Nevirapine: X-Ray structures and thermal decomposition of representative solvates. *Cryst. Growth Des.*, 8, 17-23.

Chadha, R., Arora, P., Saini, A., Jain, D.V.S. (2010) Solvated Crystalline Forms of Nevirapine: Thermoanalytical and Spectroscopic Studies . *AAPS PharmSciTech.*, 11, 3, 1328-1339.

Chemburkar, S. R., Bauer, J., Deming, K., Spiwek, H., Patel, K., Morris, J., Henry, R., Spanton, S., Dziki, W., Porter, W., Quick, J., Bauer, P., Donobauer, J., Narayanan, B. A., Soldani, M., McFarland, D., McFarland K. (2000). Dealing with the impact of ritonavir polymorphs on the late stages of bulk drug process development. *Org. Process res.* deve., 4, 413-417.

Cuffini, S., Howie, R. A., Tiekink, E. R. T., Wardelld, J. L., Wardelle, S. M. S. V. (2009). (S)-6-Chloro-4-cyclopropylethynyl-4-trifluoromethyl-1H-3,1-benzoxazin-2(4H)-one. *Acta Crystallographica Section E, Cryst.*, E65, o3170–o3171.

Desikan, S., Parsons, Jr. R. L., Davis, W. P., Ward, J. E., Marshall, W. J., Toma, P. H. (2005). Process development challenges to accommodate a late appearing stable polymorph: A case study on the polymorphism and crystallization of a fast-track drug development compound. *Org. Process res. deve.*, 9, 933-942.

Dova, E. (2008). Polymorphic forms of efavirenz. WO 2008/108630 A1.

Gandhi, R.B., Bogardus, D.E., Bugay, D.E., Perrone, R.K., Kalpan, M.A. (2000). Pharmaceutical relationships of three solid forms of stavudine. *Int. J. Pharm.,* 201, 221-237.

Gurskaya, G.V., Bochkarev, A.V., Ahdanov, A.S., Dyatkina, N.B., Kraevskii, A.A. (1991). X-ray crystallographic study of 2′,3′-dideoxy-2′,3′-didehydrothymidine conformationally restricted termination substrate of DNA polymerases. *Mol. Biol.,* 25, 483.

Harris, R. K., Yeung, R. R., Lamont, R. B., Lancaster, R. W., Lynn, S. M., Staniforth, S. E. (1997). 'Polymorphism' in a novel anti-viral agent: Lamivudine. *J. Chem.. Soc., Perkin Trans.*, 2, 2653-2654.

Harte, W.E., Starrett, J.E., Martin, J.C., Mansuri, M.M. (1991). Structure studies of the anti-HIV agent 2′,3′-didehydro-2′,3′-dideoxythymidine (d4T). *Biochem. Biophys. Res. Comm.*, 175, 298.

Jozwiakowski, M. J., Nguyen, N-A T., Sisco, J. M., Spancake, C. W. (1996). Solubility behavior of lamivudine crystal forms in recrystallization solvents. *J. Pharm. Sci.*, 85, 193-199.

Khanduri, H. C., Panda, A. K., Kumar, Y. (2006). Processes for the preparation of polymorphs of efavirenz. WO 2006/030299 A1.

Mahapatra, S., Thakur, T. S., Joseph, S., Varughese, S., Desiraju, G. R. (2010). New Solid State Forms of the Anti-HIV Drug Efavirenz. Conformational Flexibility and High Z′ Issues New Solid State Forms of the Anti-HIV Drug Efavirenz. Conformational Flexibility and High Z′ Issues. *Cryst. Growth Des.*, 10, 7, 3191–3202.

Miller, J. M., Collman, B. M., Greene, L. R., Grant, D. J., Blackburn, A. C. (2005). Identify the stable polymorph early in drug development process. *Pharm. Dev. Technol.*, 10, 291-297.

Morissette, S.L., Soukasene, S., Levinson, D., Cima, M.J., Almarsson, O. (2003). Elucidation of crystal form diversity of the HIV protease inhibitor ritonavir by high-throughput crystallization. *Applied physical sciences*, 100, 2181-2184.

Pereira, B. G., Fonte-Boa, F. D., Resende, J.A.L.C., Pinheiro, C.B., Fernandes, N.G., Yoshida, M.I., Vianna-Soares, C.D. (2007). Pseudopolymorphs and intrinsic dissolution of Nevirapine. *Cryst. Growth Des.*, 7, 2016-2023.

Radesca, L., Maurin, M., Rabel, S., Moore, J. (1999). Crystalline efavirenz. WO 99/64405.

Radesca, L., Maurin, M., Rabel, S., Moore, J. (2004). Crystalline efavirenz. US 6,673,372 B1.

Ravikumar, K., Sridhar, B. (2009). Molecular and crystal structure of efavirenz, a potent and specific inhibitor ofHIV-1 reverse transcriptase, and its monohydrate. *Mol. Cryst. Liq. Crystal*, 515, 190-198.

Reddy, B.P., Rathnakar, K., Reddy, R.R., Reddy, D.M., Reddy, K.S.C. (2006). Novel polymorphs of efavirenz. US 2006/0235008.

Reguri, B.R., Chakka, R. (2005). Novel crystalline forms of 11-cyclopropyl-5,11-dihydro-4-methyl-6H-dipyrido[3,2-b:2',3'-e][1,4]diazepin-6-one (Nevirapine). United States Patent 0059653A1.

Reguri, B.R., Chakka, R. (2006). Crystalline forms of Nevirapine. United States Patent 0183738 A1.

Sarkar, M., Perumal, O.P., Panchagnula, R. (2008). Solid-state characterization of nevirapine. *J. Pharm. Sci.*, 70, 619-630.

Sharma, R., Bhushan, H.K., Aryan, R.C., Singh, N., Pandya, B., Kumar, Y. (2006). Polymorphic forms of efavirenz and processes for their preparation. WO 2006/040643 A2.

Singh, G.P., Srivastava, D., Saini, M.B., Upadhyay, P.R. (2007). A novel crystalline form of lamivudine. WO 2007/119248 A1.

Stieger, N., Liebenberg, W., Wessels, J.C., Samsodien, H. Caira, M. R. (2010). Channel inclusion of primary alcohols in isostructural solvates of the antiretroviral nevirapine: an X-ray and thermal analysis study. *Stru. Chem.*, 21, 771-777.

Permissions

The contributors of this book come from diverse backgrounds, making this book a truly international effort. This book will bring forth new frontiers with its revolutionizing research information and detailed analysis of the nascent developments around the world.

We would like to thank Prof. Krzysztof Sztwiertnia, for lending his expertise to make the book truly unique. He has played a crucial role in the development of this book. Without his invaluable contribution this book wouldn't have been possible. He has made vital efforts to compile up to date information on the varied aspects of this subject to make this book a valuable addition to the collection of many professionals and students.

This book was conceptualized with the vision of imparting up-to-date information and advanced data in this field. To ensure the same, a matchless editorial board was set up. Every individual on the board went through rigorous rounds of assessment to prove their worth. After which they invested a large part of their time researching and compiling the most relevant data for our readers. Conferences and sessions were held from time to time between the editorial board and the contributing authors to present the data in the most comprehensible form. The editorial team has worked tirelessly to provide valuable and valid information to help people across the globe.

Every chapter published in this book has been scrutinized by our experts. Their significance has been extensively debated. The topics covered herein carry significant findings which will fuel the growth of the discipline. They may even be implemented as practical applications or may be referred to as a beginning point for another development. Chapters in this book were first published by InTech; hereby published with permission under the Creative Commons Attribution License or equivalent.

The editorial board has been involved in producing this book since its inception. They have spent rigorous hours researching and exploring the diverse topics which have resulted in the successful publishing of this book. They have passed on their knowledge of decades through this book. To expedite this challenging task, the publisher supported the team at every step. A small team of assistant editors was also appointed to further simplify the editing procedure and attain best results for the readers.

Our editorial team has been hand-picked from every corner of the world. Their multi-ethnicity adds dynamic inputs to the discussions which result in innovative outcomes. These outcomes are then further discussed with the researchers and contributors who give their valuable feedback and opinion regarding the same. The feedback is then collaborated with the researches and they are edited in a comprehensive manner to aid the understanding of the subject.

Apart from the editorial board, the designing team has also invested a significant amount of their time in understanding the subject and creating the most relevant covers. They scrutinized every image to scout for the most suitable representation of the subject and create an appropriate cover for the book.

The publishing team has been involved in this book since its early stages. They were actively engaged in every process, be it collecting the data, connecting with the contributors or procuring relevant information. The team has been an ardent support to the editorial, designing and production team. Their endless efforts to recruit the best for this project, has resulted in the accomplishment of this book. They are a veteran in the field of academics and their pool of knowledge is as vast as their experience in printing. Their expertise and guidance has proved useful at every step. Their uncompromising quality standards have made this book an exceptional effort. Their encouragement from time to time has been an inspiration for everyone.

The publisher and the editorial board hope that this book will prove to be a valuable piece of knowledge for researchers, students, practitioners and scholars across the globe.

List of Contributors

Guanghui Li, Tao Jiang, Yuanbo Zhang and Zhaokun Tang
Department of Ferrous Metallurgy, Central South University Changsha, Hunan, China

Ichiko Shimizu
Department of Earth and Planetary Science, University of Tokyo, Tokyo, Japan

Ioan Coriolan Balintoni and Constantin Balica
"Babeş-Bolyai" University, Cluj-Napoca, Romania

Pingguang Xu
Japan Atomic Energy Agency, Tokai, Ibaraki, Japan

Yo Tomota
Ibaraki University, Hitachi, Ibaraki, Japan

Rossano Lang
Instituto de Física Gleb Wataghin - UNICAMP, Campinas, SP, Brazil
Instituto de Física - UFRGS, Porto Alegre, RS, Brazil

Alan de Menezes, Eliermes Meneses and Lisandro Cardoso
Instituto de Física Gleb Wataghin - UNICAMP, Campinas, SP, Brazil

Adenilson dos Santos
CCSST, Universidade Federal do Maranhão, Imperatriz, MA, Brazil

Shay Reboh
Groupe nMat, CEMES-CNRS, Toulouse, France
Instituto de Física - UFRGS, Porto Alegre, RS, Brazil

Livio Amaral
Instituto de Física - UFRGS, Porto Alegre, RS, Brazil

Kazimierz J. Ducki
Silesian University of Technology, Poland

R.L. Brodskaya and Yu B. Marin
Saint-Petersburg State Mining University, Saint- Petersburg, Russia

Yousef Javadzadeh, Sanaz Hamedeyazdan and Solmaz Asnaashari
Biotechnology Research Center and Faculty of Pharmacy, Tabriz University of Medical Sciences, Iran

Valérie Dupray
Université de Rouen, France

Renu Chadha, Poonam Arora, Anupam Saini and Swati Bhandari
University Institute of Pharmaceutical Sciences, Punjab University, Chandigarh, India

Printed in the USA
CPSIA information can be obtained
at www.ICGtesting.com
JSHW011414221024
72173JS00004B/539

9 781632 381644